The Event in Science, History, Philosophy and Art

Yeshayahu Ben-Aharon

"The Event in Science, History, Philosophy and Art." ISBN 978-1-60264-881-4 (softcover); 978-1-60264-882-1 (electronic version).

Published 2011 by Virtualbookworm.com Publishing Inc., P.O. Box 9949, College Station, TX , 77842, US. Edited and indexed by Scott E. Hicks.

Manufactured in the United States of America.

"For the Son of Man in his day will be like the lightning which flashes and lights up the sky from one end to the other."

(Luke 17: 24)

Table of Contents

Preface

IN MY EARLY TWENTIES I experienced for the first time the fact that we are becoming co-creators in an astounding and creative cosmic evolution. I realized also to what extent this experience changes our understanding of our becoming and its role in the universe. In my books, *The Spiritual Event of the 20th Century, The New Experience of the Supersensible*, and *America's Global Responsibility*, I described this experience from the point of view of the 20th century. In recent years I have been speaking and writing about the coming and already strongly experienced event of the 21st century. In the present book I simply call it "the event."

However, the creative and uplifting energy of the event can be captured and utilized also to serve other goals. Everything positive, beautiful and valuable in our life can be also misused and reversed into its very opposite. The science and technology, that express some of our age's most creative and valuable human capacities and achievements, can be harnessed to serve one sided, narrow, and cold conceptions of nature, human nature, and the cosmos when its plans for the future of human transformation and redemption are based on purely technological, mechanistic grounds.

I have no doubt that the materialistic scientific-technological utopia will be realized, one way or another. After all, it has been well underway for centuries and it was, is and will continue to be supported by the brightest and richest. We must understand that it will demonstrate in a practical way, a complete technological mastery and transformation, indeed, transubstantiation, of everything we still consider today to be "human," including all the biological-genetic forces that determine our physical evolution, as well as all our intellectual capacities. All that we came to be, in so far as we are physically-organically embodied beings, will sooner or later be mastered by super-human technology, intelligence and creativity. Humans will produce AI machines that will make them physically immortal and infinitely clever, that is, humans are becoming superhuman, or subhuman, depending on our point of view.

One thing is therefore quite certain: the future of the human as we knew it is truly coming to an end. The only question is: what sort of an end is it going to be? Will the end of our purely natural evolution become a technological nightmare ruled forever by AI zombies **OR** will it become the beginning of our most creative, humanly most satisfying and exhilarating, cosmic adventure ever?

Introduction

PERHAPS ONE OF THE GREATEST RIDDLES of human nature is this: today we are discovering that we are really beings of *becoming* inhabiting a universe of becoming. So why are we always looking for eternal un-becoming? Could it be that, paradoxically, deep inside we know who we truly are, but are afraid of ourselves? Can it be that precisely because we are beings made of the stuff of becoming and change, we search frantically for solace and comfort in inertia, rest and even sleep, to save us from growth and change? Is this the reason for the most ancient of human drives, the desperate search for fundamental, unchangeable eternity?

This primordial, solid, eternally unchangeable foundation (be it substance, ground, first cause or God) was reformulated in the 17th century. It was named 'matter' and a host of laws of conservation were soon invoked to secure and uphold this support and to hallow it with the unchangeable eternity for which we sought. But at the same time, is not most of modern science and mathematics since the 1920s and '30s also a radical revolt against this new dogmatic materialist, positivist and scientific Theology?

But what if matter, ground and existence is nothing solid, fixed, dead forever, but a living and creative ***event***? If this is so, it would mean that "the real stuff" out of which it is all made is the stuff of becoming, evolution, and creative emergence. Entering the event means therefore that we enter the vortex of real and creative life; we are invited to become co-creators in the drama of evolution, and not only detached external spectators. Perhaps this is what

Novalis meant when he said that God is the Ur-paradox for all beings, that when the Angels contemplate God they don't get final answers from Him to their questions, but are kindly invited to penetrate ever deeper to the mystery of creative becoming, unlike humans that still look up to God to find final, eternal, immovable answers. Can it be, thus, that what is called "God" isn't the eternally fixed place of conservation of matter, identity, or essence, but the source of ever evolving transformation?

If black holes are real, and if unperceived dark matter and unknown dark energy make up most of our universe, then conservation of all sorts, be it of matter or energy, space or time, memory or idea or God, is replaced by creativity, wonder and change. We discover through the creative event of our time, a new source of emerging life, and our sciences, thinking and arts express it in radically new ways.

On March 14-19, 1979, the most distinguished physicists in the world gathered together at the Institute for Advanced Study in Princeton to celebrate Einstein's centennial (1879-1979). Roy Lisker published an account of the event *In Memoriam Einstein: The Einstein Centennial Symposium.* This meeting brought together, as never before, names such as *John Archibald Wheeler, Paul Dirac, Claudio Teitelboim Weitzman, Adolf Grunbaum, Subrahmanyan Chandrasekhar, Martin Rees, Stephen Hawking, Roger Penrose, Eugene Wigner, Isidor Rabi, Tullio Regge, Steven Weinberg, Abraham Pais, Thomas Kuhn, Yuval Ne'eman, Res Jost, Peter Gabriel Bergmann, Dennis William Siahou Sciama* and many others.

From the rich proceedings one anecdotal detail may be mentioned, because it demonstrates this turning point with fine precision. It is taken from

Yuval Ne'eman's debate with Hawking after his discussion of black holes, revolving around the very problem of *conservation*. The reason for the debate was Hawking's indication that one may have to:

> Abandon a fundamental principle of particle physics, the *Conservation of Baryon Number*: the quantitative difference between matter and anti-matter is an invariant in all interactions and processes of decay. *Yuval Ne'eman* asked if there might be some way of interpreting the equations to save baryon number conservation.

To which Hawking so aptly replied:

> *I find it interesting that people have such an emotional attachment to baryon conservation. This may be because most people do not believe in eternal life. They would like to hope that the particles which make up their bodies would live forever.*

Ne'eman answers:

> *We owe a lot to baryon number. We owe our existence to the conservation of baryon number. Otherwise we would be floating in the universe as $E=mc^2$!*

Liskar continues to report:

> Hawking's jest crystallizes the very style of contemporary physics. Scientists have been condemned throughout modern history for mocking the existence of a soul. Now it appears that they have as little use for matter! The viability of the law of baryon conservation

> seems to depend upon whom one is talking with at a particular moment.
>
> (Lisker, pp. 65-67; 1980; www.fermentmagazine.org)

Perhaps we may not immediately start to float in the universe as $E=mc^2$, but if we do give up our psychological (and often unconsciously theological) attachment to the concept of *conservation*, our thinking may expand and accelerate in approaching the event horizon of our age, confronting reality as it is, and not as we wish it to be.

As we shall presently see in the first chapter of the book, this is exactly what contemporary science teaches us. When it comes to overcoming matter and gravity, time and space, in order to float in the universe as $E=mc^2$, let us take our counsel from creatures whose very being and existence is made *of* and *for* the elements of warmth, air and light: flying insects. Naturally, it was long assumed that flying insects speed up their movement similar to the manner in which, say, horses do it, when they change from walk to trot to canter. That is, in sense-perceptible, physical and linear, continuously advancing adjustments of essentially identical gestures of movement. Recent investigation has revealed the very opposite of this natural assumption. Insects, in fact, use all known methods of flight available in order to execute each unit of speeding up. They do not advance in a linear fashion at all; but, surprisingly, they use all possible maneuvers. If a human could move like that, instead of just walking and then walking faster and then running in a straight forward manner, she would hop, skip, jump, cartwheel and back flip, and in no fixed order at that. We don't move this way in the physical world, but could it be the case that this is actually how we are able to create our new ideas, inventions and works of art?

The remarkable fact is that insects don't execute each movement separately, and then move in a linear fashion from one movement to the next. Between one wing stroke and another, when flight speed changes, insects execute the whole possible range of movements with almost infinite speed. In other words, they transcend time and speed altogether, and control their flight as a whole from a virtual plane, beyond time and space (I follow Bergson and Deleuze's use of the term 'virtual' to mean the infinite potential of a thing).

We can actually see and photograph with high speed cameras, how infinite wholeness is taking hold of time and space and is realized therein as *event*; an infinitely fast survey which is not really a physical acceleration connecting, one after the other, all flight possibilities, but an untimely and unspatial grasp of the *infinite wholeness of flying*. What appears in time and space—the infinite speed of the actualization of all possible flight possibilities—is not directed "from below," that is, from space and time, but "from above," from the infinite virtual plane that contains (in one survey) all these possibilities, and much more. The insect's flight then displays almost tangibly in time and space the actualization of an infinite virtual plane. It is demonstrating what we call "virtual actualization," the wholeness of a being or an event actualizing itself, incarnating in time and space, yet not exhausting itself.

What we glimpse here, what lights through the executed acceleration, or better the *self-actualizing virtual wholeness* of the insect's flight possibilities, is the true virtual home of butterflies, and, if we study *ourselves* more closely, we find that it is also the home of the creative activity of humans. May it not be discovered, the closer we draw to the "event horizon" of our time, that also humans—as beings of perpetual becoming—share with their thinking and consciousness the same world in which infinity is an active, virtual force?

And how otherwise could true creativity be explained, how could it ever become possible? What makes the new, the creative, possible, if not our share in *infinite wholeness,* and our ability to bring it down to earth, actualize it and individualize it in space and time, matter and life, in human relations and in works of art? Modern science, philosophy and art teach us this new art of creativity. They are creatively thinking and actualizing the event. This book describes some of their creative accomplishments.

Charles Tomlinson described the event this way:

The breath of circumstance
is warm, a greeting...
and under each death, a birth.

("One World," in the second
number of The Resuscitator,
1964)

Chapter One: The Event in Science

Who possesses science and art,
Possesses religion as well;
Who possesses the first two not,
O grant him religion.
(Goethe)

Clouds are not spheres, mountains are not cones, coastlines are not circles, and bark is not smooth, nor does lightning travel in a straight line.

(Benoit Mandelbrot, *Fractal Geometry of Nature*)

The fact that we can bring together the truth of scientists and poets is.... Already a certain proof that we can bridge the chasm between the 'two cultures,' and that we have a possibility for a new dialogue... we are beginning to envision a new unity, realized by a non-totalitarian science, [a unity] in which we don't strive to reduce one realm to the other...

(Ilya Prigogine, *La Nouvelle Alliance*)

In the depth of the human being there is probably a common source... for religious belief, cosmic observation, scientific analysis and aesthetic experience. And in this common framework is grounded a deep relationship between science, art and religion, through which the human creative power comes to full expression.

(Katzir-Katchalsky, *In the Crucible of Scientific Revolution*)

Part One: A New Synthesis

A NEW SCIENTIFIC AND CULTURAL PARADIGM has emerged since the second half of the 20th century, as a result of a renewed encounter between the natural sciences and the deeper questions, problems and riddles of human becoming. This is part of the general social and cultural change taking place since the '60s of the last century. The new paradigm in the sciences is in itself one aspect of an overall change of human consciousness that continues in various forms today. In a continuously expanding periphery, here and there already in some mainstream centers, natural and human sciences talk more and more in a language that gives expression to these changes in our perception and understanding of the meaning of our life. In science, this change is sometimes called "the second revolution in natural science," relative to the first revolution, in which classical physics was founded in the 17th century.

The greatness of the mechanics of Kepler, Galileo, Newton and Laplace is based on the fact that it can solve basic problems of the simplest behavior and movement of physical bodies. However, this requires the presupposition of an **abstract space and a neutral time** and experimental conditions in which many of factors of real life, such as friction, chance, and entropy have been removed. Mechanics deals with systems of bodies that can be quantified mathematically, and predicts with impressive accuracy the behavior and movements of such bodies. This caused Laplace, a great mathematician and a believer in Newtonian physics, to declare that:

> An intelligence that, at a given instant, could comprehend all the forces by which nature is animated and the respective situation of the beings that make it up...would encompass in the same formula the movements of the greatest bodies of the universe and those of the lightest atoms. For such an intelligence nothing would be uncertain, and the future, like the past, would be open to its eyes.
>
> (Pierre-Simon Laplace, *Philosophical Essay on Probabilities* (1814))

In other words, mechanical systems are so precise and predictable because they are **ab**-or **sub-tracted** from the real world. They are strictly firewalled and disconnected from the normal physical flow of matter, energy, life and cognition, and in this man-made, non-real, ideal world, such an absolute determinism is really successful. The behavior of objects placed in this ideal mechanical world can be predicted prior to their movement and the arrow of time can be neutralized. Of course, since the 20s and 30s of the last century, this view was challenged through quantum and atomic physics as well as some aspects of the theory of relativity, but this confrontation took place only in the esoteric world of sub-atomic particles. The next step of the second scientific revolution has lifted this change from the invisible world and placed it in the open and the real, visible, world of natural and human evolution and becoming.

Today we are much more open to admitting that in the strictly Newtonian worldview there is no place for creative evolution, development and new emergence and becoming. Now the paradox is that since the 19th century, the most fundamental evolution of science is precisely this: that evolution as such was discovered and since then it has become one of science's most ardently pursued, interesting and, indeed, creative fields. Ben-Jacob, a leading scientist and thinker of the second scientific revolution, referred to this paradox in this way:

> One of the big ironies of scientific development in the 20th century is that its burst of creativity helped establish the hegemony of a paradigm that regards creativity as an illusion. The independent discovery of the structure of DNA (Universal Genetic Code), the introduction of Chomsky's notion about human languages (Universal Grammar) and the launching of electronic computers (Turing Universal Machines), all occurring during the 1950s, later merged and together established the dominance of reductionism.
>
> (Eshel Ben-Jacob and Yoash Shapira, *Meaning-Based Natural Intelligence vs. Information-Based Artificial Intelligence. Tel Aviv University, c. 2004)*

But this means that precisely the four fastest developing scientific narratives of our time: evolutionary cosmology, paleontology, history and psychology, cannot find in classical physics any consistent scientific and philosophical foundations. In fact, in the eyes of classical physics, time has no direction, past and future are the same, and therefore evolution itself must be viewed as mere local illusion. In its view, evolution emerged without any possible explanation, and will disappear with no consequences.

In the new paradigm, however, matter, earth, and nature with all its creatures, the human and the universe exist in a continuous state of emergence, which means, in other words, in creative becoming and transformation. In our immediate earthly environment, to mention only one example, this becoming takes place in astoundingly complicated ecological systems in which millions of species and classes take part in synergetic and symbiotic relations mutually interconnected to each other's development. As we shall see in greater detail below, even the stronghold of the old scientific paradigm in biology called the "new synthesis" of mechanistic-analytic evolutionary theory and genetics is currently undergoing essential changes. Today it becomes increasingly clear that the selection, variation and adaptation of species are far more complex and indeed interesting than the notions of brutal self-preservation as the source of "survival of the fittest."

What is now more and more emphasized is the fact that evolution is a process of cooperative becoming in which the conventionally conceived hereditary genetics are not the sole and perhaps not even the main agent of change. The new formula for evolution and development alike (evo-devo), really tells us that we are only beginning to skim the surface of the most complex and creative, cooperative and synergetic field of development and evolution. Heredity and adaptation, selection and variation, change and metamorphosis are no longer viewed in terms of *linear, mechanical causation*, where the information is predetermined, fixed and encoded in the genome, which in turn produces the proteins that build the cell like an assembly line in a factory. Quite the contrary; scientists today are fascinated by the inherent (as well as inherited!) **potential** for change, chance, mutation, and variation that penetrate all the structures and functions involved in development and evolution. Potential for change exists in the DNA itself, in the cell's cytoplasm and environment, in the planetary ecological system, and indeed in the cosmos as a whole. This potential for change appears now to be of far greater scientific importance than the actualized genetic information. The genome may well store the information of what has evolved **in the past**, but it must also be conceived in such a way that shows how it is open **for the future**, how it integrates the next emerging and creative stage of evolution.

For this reason, the place of human becoming in nature changes as well. We become less strangers in this new scientific picture of the universe, and begin to feel more *at home in the universe* (to use Stuart Kauffman's beautiful book title) because the universe itself becomes more "human like" and less a result of deterministic mechanics or *arbitrary* chance. It becomes a place in which becoming is for the first time **scientifically** thinkable, and everything natural, cosmic and human becomes a far more complex and fascinating puzzle on all levels and dimensions. The physical universe, the kingdoms of nature, the human, and the complex connections between them, seem today more like the emergent process of a great work of art, crowned with intelligence, imagination and rich with unlimited creative potentials, than mechanistic clockwork or what truly amounts to the same, a finished product of a divine "intelligent designer." (This really begs us to rephrase and enlarge Blake's protest against Newton to include the new oldest intruder: "May God us keep from single vision and Newton's sleep" and intelligent designer's religious grip!)

In fact, many scientists today partake in the creation of an intellectual, philosophical, cultural, and creative narrative in which surprising and new interdisciplinary meetings take place, which reveal a previously unrecognizable world of interrelationships between all fields of knowledge and creativity. One can sense, if one listens carefully to such discourses, that there is a new sense of wonder, readiness and openness for new ideas that begins to seep even into some strongholds of what used to be purely analytic and reductionist science. This is also the reason why the second scientific revolution, if it is viewed from a broad cultural prospective, may be seen as part of a greater revolution of consciousness itself, and therefore can justify what I may term "the creative adventure of our times," which is clearly not only interdisciplinary in the academic sense, but encompasses society and culture as a whole. It brings together representatives of many sectors of our society, including scientists, artists, thinkers and socially creative people.

The new interconnections created today in the meeting of philosophers, scientists or artists with historical and social events bring a sense of the possibility for a deepening of the meaning of human life. In all civilizations, at all times and in all places, such meetings have created the real atmosphere of creative adventure, the power which drives and enlivens personal, societal and spiritual becoming and growth. It seems that already in the first decades of the 21st century there is a possibility that such meetings can become a source of inspiration for an emerging new social sensitivity, supporting a new politics and economy, and striving to actualize a society different from the one we knew in the 20th century.

Some of the major trends of change created and synthesized together in the course of the twentieth century are composed by the connection of three concurrent transformations of human consciousness:

1. The second scientific revolution in the natural and human sciences.

2. Poststructuralism in philosophy, humanities and the arts.

3. Research and development of altered states of consciousness and cognition.

The synthesis of these three transformations creates radically new concepts and practices in all of the fields with which we have been living since the end of the last century and the beginning of the present. Some central new concepts that belong to this creative synthesis are discussed in what follows.

Creative Emergence and Becoming

If fixity and inertia were the basic characteristics of the older conceptions of earth, stars and cosmos, our universe has become a stage of constant emergence, happenings, and becoming on all levels. The new physics, astronomy and astrophysics describe the universe not only as full of information in its present structures, but also as a stream of becoming endowed with unlimited creative and emergent potentials. It is portrayed as if its very existence expresses itself in continued becoming and the formation of novelties, with chaos, order, complexity, and creativity as its main tools. Our concepts of *matter* and the material world have changed in the same manner. Matter is beginning to be understood as a platform of continued spontaneous becoming, which allows for the emergence of life and consciousness in the material world. Life, consciousness, and self-consciousness are no longer considered strange and even contradictory to the mechanical laws of physical science, but rather an expression of the creative potentials embedded in the universe. The evolution of the earth, life, and consciousness may not be just a meaningless event in a meaningless universe, neither the creation of the old external God now renamed "intelligent designer," but a creative adventure in which we may have a role to play which far exceeds our present limited imagination. A hint in this direction is implied in the fact that some of the fundamental laws of physics are described today as more and more similar to the creative processes that we discover in our own lives and minds. We begin to feel at home again in the universe because we have changed the way in which we see it, and this in turn changes the way we see ourselves.

The Web of Life

As Fritjof Capra suggested in his book, *The Web of Life: a New Scientific Understanding of Living Systems*, the new insights in biology, ecology, and earth sciences describe the whole Earth as a living entity. Such a conception was of course always there in ancient mythical traditions all over the earth. The point is that today it has reemerged, but now as the result of our exact scientific studies. The lithosphere, biosphere, and atmosphere, are integrated together with dynamic life cycles, wisdom, beauty, and creativity. The fabric of planetary life is described today as the circulation and respiration of "Gaia," the living earth. In her fabric we find the cycling of elements, the forming of continents, seas, climates and all the kingdoms of nature, interrelated as an essentially living organism. This web of life is the vehicle and matrix that carries and supports the development of animal and human consciousness in the material world.

Humanity is evolution becoming self-conscious

Humanity in general, and every individual in it, represents in the universe a place where evolution becomes conscious of itself. Evolution is becoming conscious of itself in and through the human. Does humanity express, in a human way, through thought, language, art, invention, the same universal forces from which the universe is composed? Do we discover ourselves when we discover the new laws and processes in nature around us and in the universe above us? It may turn out that humanity is the first place in the universe, as we know it, at which the infinitely creative cosmic formative forces become conscious of themselves. This will have changed entirely not only our picture of the world, but also our picture of us in this world. We may be approaching a time in which the human will no longer consider the earth and itself as a mere meaningless speck of dust in a totally alien universe, but as a self-conscious representative of, and co-creator with, the basic evolutionary forces of the cosmos.

The human as the growing edge of cosmic evolution

If this becomes true, then the creations of humanity, in culture, society, science, and technology, will be understood as expressing a continuation of the universe's evolution. Human intelligence will be experienced as an expression of cosmic intelligence, human creativity as part of cosmic creativity; and the evolution of life and consciousness may be seen as a direct continuation, variation, and metamorphosis of cosmic evolution. In this sense it will be possible to experience through the development of the cognitive and moral potential active in humanity, that cosmic evolution continues to renew itself, emerging with previously unexpected horizons. Can we begin to imagine that human creativity, which is changing the earth at such a fast pace, is part of the creative forces of the universe? Will humanity be seen in the future as a concentrated essence, a seed, that on one hand, inherits the past treasures of cosmic and planetary evolution and on the other, harbors new, unimaginable, future cosmic possibilities? Is the becoming process of the earth, nature and humanity in reality an emergent seed, teeming with future creative potentials of worlds yet to become?

Individuation and future cosmic integration

There is also a commonly shared assumption among divergent streams of thought in the humanities, that *individuation* is a key factor to be reckoned with, if we wish to understand the great change of consciousness taking place in the new era. But in some fields a somewhat pessimistic tone is heard when the psychological and social results of this process are described. It is often suggested that individuation is responsible for our increased feeling of estrangement from nature and the enhanced sense of alienation from other human beings. However, individuation may show a more beneficial side if we place it in an evolutionary context, especially so, if we consider that without it no *human freedom* would be at all possible.

Of course, there is no denying the tragedies brought about by the misuse of this freedom in modern life; however, there is also a cautious sense of hope arising from the same sense of individual freedom, a search to find deeper connections between one person and another, between earth and cosmos. Can we imagine for a moment that individuation could be a positive, even necessary, starting point, not condemning us to suffer eternal isolation and estrangement, but calling us to develop new forces of creative integration? Can individuation be transformed into a motivating power for new human-social and cosmic integration in which the newly discovered human self will not be extinguished, but rather *expanded and enriched*? Is it possible that becoming a real free individual would not just be the end of the ancient integration between the human, society, and universe, but also the starting point of a novel and free integration?

The Two Cultures Revisited

In the lectures that the physicist *C. P. Snow* gave in 1959 under the title *The Two Cultures and the Scientific Revolution* he claimed that our culture is split in two separate cultures. On one side stand the intellectuals from the humanities and literary studies, in charge of the majority of the academic journals and literary reviews in the popular press, who therefore heavily define culture among academic and educated circles.

On the other side stand the scientists who really change the nature of our perception of reality, but who are mostly shut in academic ivory towers, and express themselves with esoteric jargon far removed from the mundane world. The intellectuals, Snow claimed, are at fault for the cultural aversion to science because they have not endeavored to understand, critique, or join in creative partnerships with the developers of the ongoing scientific revolution.

According to his opinion, this revolution, especially in physics, beginning in the thirties of the 20th century, created new fundamentals for our modern worldview, and in essence a basis for a second scientific revolution. Snow mentions among others, the astronomer *Edwin Hubble*, the mathematician *John Von Neumann*, the cyberneticist *Norbert Weiner* and the physicists *Niels Bohr, Albert Einstein*, and *Werner Heisenberg,* who discovered and created, each in his own field, not only new scientific concepts but also concepts with general human value and general cultural and spiritual meaning. However, scientists also contributed to this division. Even though physicists made in the first half of the twentieth century the greatest contribution to a new world view, they did not receive widespread support from their colleagues in other fields. In 1959, Snow did not realize how fast the situation was going to change over the next years. Even as he spoke, the fields of cybernetics, general systems theory, open systems physics, ecology, and existential and humanistic psychology, sociology, and history were combining to create a wider interdisciplinary basis for the new paradigm, based on an expanding second scientific revolution which began with quantum physics and the theory of relativity.

In 1963, in the second edition of The Two Cultures, Snow added a new chapter in which he establishes that a new culture, a "third culture," will emerge in the coming decades, as a result of the closing of the gaps between the intellectuals and the scientists. The intellectuals will begin to understand the language of scientists and vice versa. Many people will feel the need for an enhanced dialogue between the two cultures, because the representatives of literature, philosophy, and the arts, will discover that the changes which are taking place in the sciences bring them closer to the creative source of humanity. In effect Snow's prediction has come true in a greater sense. As the twentieth century came to a close, new scientific concepts became the subject of widespread literature and multimedia, and their influence on wider circles of culture and society is greater now than their influence on intellectuals.

What took place beginning in the 1950s and '60s, which contributed, if not to the closing of the divide between the two cultures, at least to an increased interpenetration? Are we really justified in speaking about an Event of new emergence that has a broad cultural significance? As we will demonstrate below, it is possible to answer this question positively. Beginning from the mid 1950s and even more so since the '60s, monumental changes have occurred in all fields of thought, from scientific research to the arts and cognitive experimentation. It seems that human awareness in general began to undergo extensive changes, which are expressed in all realms of life. The '60s have by now become a well-established concept in social and historical research. Since then sociologists, anthropologists and historians have spoken of these years as the beginning of a new age or new era in which wholly new horizons of thinking, creativity and life were opening.

The Structure of Scientific Revolutions

This general cultural and social change shows itself clearly in the change in a ruling scientific paradigm. This was the meaning of a new concept that Thomas Kuhn introduced—exactly at the right time—into the study of the history and sociology of science, in his classic book *The Structure of Scientific Revolutions* (1962). Kuhn discovered that if we view scientific research and life in its entirety, we realize that it takes place along **two** axes, one horizontal and one vertical. The vertical axis signifies the drive for discovery and creative thought. It is this trajectory which blazes the way and opens unpredicted and startling new horizons. The horizontal axis on the other hand is involved in taking control of and securing the newly found scientific and intellectual territory. According to Kuhn, the evolutionary leaps which occur in science take place in the following way. Breaking through to a new paradigm is always at odds with the old established and entrenched paradigm. It is a hard fight that takes a very long time. But once a breakthrough has taken place, it is harnessed immediately to the grounding of a new dogmatic paradigm. That is, once it has succeeded in its fight against the mainstream, the days of its emergence are short, as sooner or later the same forces that tried to prevent its emergence are now appropriating it to the same goals of conservation. The scientific community reabsorbs the truly new back into established norms founded on the current cultural, social, and political dogmas and worldviews.

Transforming, shifting, or breaking old paradigms and growing new ones happen vertically, while controlling, using, and normalizing the achievements of the new happen horizontally. Conventional science operates on the horizontal axis. It finds its unique role and social responsibility in establishing the validity of the new paradigm, and displaying its correctness. The purpose of such research is to create a smooth logical surface that is devoid of contradictions and further problems and riddles. Kuhn showed that this is often accomplished by *ignoring* paradoxes, contradictions, and black holes in theory and practice alike. In contrast, on the vertical axis, an open ended search *highlights* the contradictions, problems, and black holes in the ruling paradigm, with the assumption that the unseen discoveries of the future can be found in just these places. The activities, which take place along this axis, demand an open mind for the unknown and unexplained, skepticism regarding the conventional wisdom, and a bold sense of adventure.

A revolution in paradigms occurs as soon as there is a critical mass of unresolved questions or novel approaches that shed light on old solutions and changes them into new questions. And this fresh avenue must be ingrained in the (relatively) open minds of a new generation of researchers. Let us present one significant example from the fields of physics and chemistry which demonstrates to what extent physical science has become a place of *becoming*, a powerfully fertile seed-bed of human and spiritual forces of the future.

Ilya Prigogine

In the year 1955, a Russian born, Belgian physicist and chemist, *Ilya Prigogine*, published a study of **open systems** in his book *Thermodynamics of Irreversible Processes.* This study earned him the Nobel Prize in chemistry in 1977. In the book *Order out of Chaos* (1984), which he wrote together with science historian *Isabelle Stengers,* Prigogine argued that the time has come to find scientific ways to insert the concepts of time, becoming, and evolution back into the very heart of modern physics. Before Prigogine's research, *time* and *becoming* were largely considered to be rather embarrassing stepbrothers who were not compatible with classical mechanistic physics. But Prigogine showed convincingly that classical physics deals only with **closed systems** which are actually very rare in nature and the universe, and that classical physics took the exception for the rule and instituted a whole scientific paradigm on this exception. But it is clear that physical, bio-chemical, and cosmological processes operate, in the actual world, through *open systems,* where the emergence of new and unforeseen facts is not the exception, but the rule.

Look at time for example. It flows stubbornly in an irreversible direction from the past to the present and opens up for future becoming. A material object in the real world, like a cup, is an open system, produced by means of an external investment of energy and knowledge, and it breaks down irreversibly. A plant, like any other living organism, is an open system for the flow of cosmic energy and it grows, develops and multiplies in real time from the past to the present and future, but it also moves backward in time, because it "returns" back to the seed and starts all over again. In closed systems, there is no difference between past and future, left or right, up or down... symmetry is absolute for all parameters. In open systems, however, irreversible processes are the rule of the game, and take place in every material and living being which progresses through birth, growth, development, decline, death, and rebirth through a new seed or innovative concept. Prigogine believed that he could prove that such systems had physical laws which could be demonstrated scientifically, analyzed mathematically and quantified adequately, and that, therefore, he could create a bridge between the two cultures, and between the sciences of matter, ecology, life and the humanities. In the original French, the book *Order out of Chaos* is actually called *La Nouvelle Alliance* (The New Alliance) and is subtitled fittingly as "The new dialogue of man with nature."

This dialogue was created after a period of 300 years of separation and estrangement between the human and nature, and therefore also between the human and itself. Both classical physical science and the humanities failed to bridge this gap until the middle of the twentieth century. Matter was conceived as life-less material, and life and consciousness were reduced to this same level. Both life and consciousness were considered to be exceptions or meaningless deviations from the earthly and cosmic laws of physics, and "real nature" was believed to be mechanical and inert. When irreversibility was at all admitted, it was only to allow universal entropy to hold an absolute and deeply pessimistic sway. A 'suicidal' universe was portrayed, which irrevocably followed the second law of thermodynamics all the way to the final, entropic, heat-death of the universe. Many scholars and artists believed this no less than the physicists at the end of the 19th and the beginning of the 20th centuries. Prigogine and his colleagues, succeeded in changing these beliefs, which had been almost hardened into universally accepted scientific dogma.

According to Prigogine, the book *Order out of Chaos* was meant to reverse this scientific world conception. The human (of real matter, life, consciousness, creativity, etc.) is no longer a "freak" in nature and in the universe. On the contrary, nature, matter, and universe appear increasingly "human like." He wished to demonstrate that Monism, or Holism, is possible also from a non-reductionist perspective: instead of making everything one and the same by reducing them to the lowest level of an imaginary material-mechanical world, we are now in the position to lift up matter itself into the higher levels of life, consciousness and creativity. And one of the major breakthroughs that enable this transformation (or reversal of the classical, original, mechanist reversal) is the new ability to offer fully scientific explanations of the experience and concept of real time, without reducing it to the classical concept of mechanical time:

> This is the contrast of our view and the classical view of Jacques Monod, that said that the old alliance between humanity and nature is broken, and this is the reason why we are alone in the universe. His conception is based on the assumption that classical physics is reversible in time [mechanical], but all life and human phenomena aren't reversible. I think that what makes the difference between the two cultures is the concept of time: there are real events in history, but not in classical

> physics. Therefore, a new concept of irreversible time in the sciences of matter enables us to create a bridge over the gap between the two cultures. In other words, when we bring the concept of time into the description of the universe, it allows us to see the human again as an inseparable part of the universe.

If the concept of creative emergence becomes the foundation of the fourfold evolutionary narrative of modern science: cosmology, evolution, history and psychology, we can begin to estimate the magnitude of the second scientific revolution. It then becomes possible to base the four narratives on new physical foundations, which support life and do not oppose it. Until now we have been trying to bridge the gap between physics, biology, evolution, history, and cosmology through the reduction of real matter, time, life and consciousness to the laws of classical mechanics and thermodynamics which exist only in artificially closed systems. Now for the first time we have the ability to effect a reversal of this reduction in all the branches of science. It means an expansion and opening of physics "upwards" to the cosmic reaches of the *Event:* the source of creative becoming and emergence of the new and unexpected, as opposed to reducing and contracting it "downwards" in order to make it compatible with the mechanical laws of classical physics.

Katzir-Katchalsky: Pioneer of a bridge between natural science and the humanities

In the author's forward to the English translation of *Order out of Chaos*, Prigogine and Stengers mention their indebtedness to many researchers, whose collective work created the basis for their study. They dedicate the book in memory of the scientists who passed prior to its publication. Among the names, alongside *Erich Jantsch, Pierre Resibois,* and *Léon Rosenfeld,* appears the name *Aaron Katchalsky* who is better known by his Hebrew name, *Aharon Katzir.*

Aharon Katzir-Katchalsky, one of the most important contributors to the bridge between the cultures, was an original thinker and scientist. Many years before the pioneering work of Ilya Prigogine and his colleagues became known, and before the popularity of the interdisciplinary ramifications of chaos theory, Katzir understood the revolutionary consequences in this field. In May of 1972, Professor Katzir led a special group of worldwide experts at MIT in a conference which, when viewed in hindsight, was a pioneering and groundbreaking event not only because of the breadth of the disciplines involved, but also with regard to its new approach to the connections between physics, chemistry, biology, and neurology, to the research in cognition, consciousnesses, and the brain.

Katzir explained to the attendees in the conference the evidence gathered through his research in *network thermodynamics,* and their applications for the understanding of complex systems in chemistry and biology. The research of Katzir and his colleagues at the Weizmann Institute (Professor Orah Kedem and others) on heat conduction through membranes — living and artificial — confirmed Prigogine's later discoveries because they could prove the role that small and sudden fluctuations played in complex and open physical and biological systems.

However Katzir did not stop here. He predicted that it could be possible that the theory of Prigogine with regard to chemical open systems, which exhibit chaotic order far from equilibrium, could offer a valuable contribution to understanding the functionality of the brain, cognition and consciousness. In proposing this, Katzir for the first time connected his and Prigogine's research, with the neurological sciences, brain research, and cognition. He blazed a path, which would influence future research, in physics, biology, neurology, and cognition for the next thirty years. Katzir outlined his vision that in the near future, there would be a need for a wholly new synthesis not only of all the sciences, but also of all social, cultural, and intellectual fields. This new synergy would serve the most monumental research ever undertaken by human thought: to understand the human being in its wholeness, as a being who combines in such a unique way the physical, biological, and neurological with emotional, cognitive, artistic and spiritual creativity.

If we rely on the research done over the past decades—so did Katzir tell the gathered scientists at the conference in 1972—we are now in a position which we could not have been beforehand in human history: to begin to scientifically, yet holistically, approach the "the ultimate question," which is: what must be the physical and *biological characteristics* of the human brain in order to enable the *operations of the human mind*? That is, Katzir believed that the new scientific advances, the new paradigm shifts, would be able to support, in a non-mechanistic, non-reductionist way, a foundation for understanding scientific, philosophic, religious, and artistic creativity. On one hand, the brain must exhibit stable, balanced, and ordered states, to allow the mind to operate on a secure basis. On the other hand, creativity in all fields, not the least in scientific inquiry itself, forces the mind to be open to the new and unforeseeable, to spontaneous developmental processes, to randomness, and to mistakes. How and where does the brain make the shifts which allow these transitions from closed systems to open ones, from order to chaos, and from simplicity and stability to higher, more complex, and less stable orders? Or how does it jump back and forth from the mechanical to the biological, from the biological to the psychological-cognitive, and from the cognitive to the creative spiritual, while remaining open to the social environment, cultural interactions, and cosmic influences?

How do these changes take place? What exactly is the physical, neurological, chemical, and psychological process, which has the ability to allow such changes, *from* the mechanical realm which is completely oblivious of time and always strives for maximal entropy and closed, balanced states of energy and information, *to* states that are precariously open, unstable, and yet linked to the surroundings? How do these open states let the inflow of energy, matter, information and meaning flow back and forth, in an active way, to develop through self-organization even greater levels of integration, complexity and order, and achieve all these remarkable feats opposing the direction of classical thermodynamics and entropy? Only by means of such open brain processes can human consciousness be what it actually is, an active and energetic creative agent in nature, who creates worlds through science, technology, arts, social relations and culture, that have never before existed and are new and unpredictable—for good or ill.

In classical physics, the atomic and subatomic world is described as if all the particles (atoms and their parts) are identical and equally stripped of any special qualities. For example, it makes absolutely no sense to ask: *what is the color of a gold atom?* Yet when we talk about complex systems, and especially living organisms, we pinpoint unique individual qualities and characteristics. What causes an immensely large group of atoms and particles to become and behave as **one united system** in which they are interconnected and interdependent, and, moreover, how do they take on unique, common qualities (a piece of gold as real substance)? What gives a system the ability to self-organize itself, to design itself, to recreate itself, to become an individual substance? What makes the infinite atomic and cellular diversity, which constitute even the smallest organism, to become a functioning whole? What makes the human being into a personality, into a singular entity? And this is definitely not, according to Katzir, a ghost in the machine; he was looking for the *scientific* basis of wholeness.

Katzir believed that if we could understand the secret of the transition from one level of organization and identity to the next, from the non-organized and chaotic, to the organized, and back to chaos, in order to achieve a higher level of organization, up to the level an individual organism, we may begin at least in a rough way to understand the functioning of the most complex natural systems, the human brain, that crowns its complexity with constituting the foundation for an experience of personal identity and social connectivity. "When does the individual begin?" asks Katzir in *The Crucible of the Scientific Revolution*. He answers, "In the living organism it begins with the ordering of the single molecules into more complex systems, for example the cell, and the individual organism is a vast and infinitely more complex combination. In the organic being the principle of individuation reaches its highest level. Now if the difference is not at the atomic level, it must be found in operation at the next higher level, the molecular level, which is created when the atoms form molecules." And then Katzir quotes from the medieval philosophic classic *The Guide to the Perplexed*: "All the parts are similar and equal, there is no difference whatsoever between them. And *the being of the whole* [the original Hebrew term used here by Maimonides, *Kibbutz,* means 'a gathering together'] is created through their interrelations." That is, when the isolated particles are gathered together, they enter into an interaction of becoming, that allows "creative emergence" of something that was not there at least explicitly before, a whole not only much greater but essentially different from its parts, a new material and new function. And it gives expression to a *singularity*, self-creating and self-reproducing in and through its greater environment.

Katzir posed the following question to the listeners, whose initial surprise at the originality of the presentation, was replaced with increasing attention. Is it possible that complex chaotic systems achieve in a spontaneous way, a state in which myriad particles raise their amount of energy and information and become one functional, singular, entity? Is it possible, as Maimonides says that in this situation the particles gather together in a *Kibbutz* formation, as if obeying a magic call, and re-assemble and re-emerge as one entity, which functions, behaves, and expresses itself as a singular self identity, be it material, biological, psychological, social or spiritual?

Many studies since the '70s show clearly that processes of this sort take place constantly in the nervous system and it could very possibly be that they are important partners in the creation of the needed physiological basis for the relations between brain, consciousness, and the mind. For example, following brain waves through an EEG shows a correlation between voluntary action and small changes in brain function. Situations of concentration, awareness, focus, and also meditation, calmness, artistic pleasure, strengthen the slowest and largest brain waves, Theta and Delta, which means that they express a global pattern of transformation in the brain. Katzir suggested, therefore, that such radical changes may cause or underlie a widening or deepening of awareness, cognition and consciousness, and may allow us to also probe the unique creative processes of the human mind.

Later at his talk at MIT, Katzir noted that Gestalt theory noticed long ago such changes and jumps in perception. Or in his words, "The transformation of the individual personality can take such sudden shapes like strikes of insights, grasping new ideas, falling in love or experiencing something like Paul's epiphany on the way to Damascus." Katzir's words found many attentive listeners at the conference. Walter J. Freeman III from the department of molecular and cellular biology at Berkeley, one of the most distinguished brain researchers today, presented a short film which demonstrated Katzir's ideas. It showed that brain waves which are created in the smell buds of rabbits function in a dissipative way, which might suggest that this perceptual part of the brain is organized as an independent yet interconnected system, and that it shows a hierarchy of semi-autonomic levels, each of which has the power to develop high levels of organization as a result of sensory, emotive, and cognitive processes. "Katchelsky was excited like a child at his birthday party," Freeman remembers. He also said that Katzir played a short but crucial role as a catalyst, in the development of the new paradigm in the understanding of the mind and its function. "Today we say that the mind is a self-organizing system. It is not an automatic or a deterministic machine. I believe that this specific insight did not emanate only from Katchelsky's vast understanding of science, but in yet deeper way it was a result of his philosophic, humanistic, and optimistic conception of the human being as a whole." (Freeman was later to implement Katzir's insight in his researches of the olfactory processes and the brain as chaotic systems. See, for example, his highly interesting work: *"How brains make chaos in order to make sense of the world,"* Christine A. Skarda and Walter J. Freeman, *Behavioral and Brain Sciences, 1987, 10, pp. 161–195.)*

Those who were present at the conference will later express a feeling, shared by many there, that they were witnesses to one of those special moments, a real "event," during which something occurs which can only happen at a great crossroad in time. Here different directions in science and thought met at an interdisciplinary intersection and began a process of mutual sharing and enrichment by means of which they planted a seed of new creative synthesis, and contributed to the creation of a new paradigm and world view. Could it be that in chaotic and dissipative systems, open to the universe and in a state dynamic and without equilibrium, a key may be found to future possibilities in the study of the mind? It was clear to everybody that the subject was worth studying and researching. Katzir was requested to organize and direct the continuation of the research in this direction and he was given the mandate to gather an international interdisciplinary team for this purpose.

The personal (Katzir's life and scientific work) and the universal-human (the development of science in our time) met at a significant spiritual crossroad. Katzir was there because he was one of the first to realize the significance and importance of this crossroad. When he returned to Israel from the conference at MIT, he was ready for a new and adventurous chapter in his research, both as a scientist and in his creative life as a whole. However, on the thirtieth of May 1972, he was murdered during a terrorist attack at Ben-Gurion airport in Tel-Aviv. Two years after his death, the proceedings of the MIT conference were published under the title *Dynamic Patterns in Brain-Cell Assemblies,* which marked from then on, a central direction in the second scientific revolution in the natural sciences and in the development of the new scientific paradigm. In the creation of an intimate connection between the research of mind and consciousness and new ideas in physics, chemistry, biology and neurology, Katzir planted an invisible and yet potent seed, which continues to play an important role in bridging the divide between the two cultures. The building of this bridge which is intended to stretch between the spiritual worlds, culture and human society, on one hand, and nature, matter, cosmos, on the other, is the great labor which is awaiting for the creative spirit of humanity in the twenty-first century.

Part Two: The Whole is the Open and It is Outside

THE NEW SCIENTIFIC VIEW of the world shows that everything is in a process of continued development and evolution. We use the term "becoming" to name this power operating in open, complex systems that exist far from equilibrium. Now the remarkable discovery of Prigogine and others is that most material systems are actually open systems and that what was called in classical physics "dead matter" exists nowhere. What happens in this matter is in reality more similar to what takes place in the living organism than in the classical mechanical systems, because it appears that not only living organisms, but that matter itself is an open system, exchanging energy, information, entropy, and negatropy with its environment. The ideal of a totally lifeless and inert matter is replaced with real world matter, and this (real, not merely ideally thought up) matter is the stuff of becoming itself.

In classical physics and science as a whole, life (and consciousness, mind as well as all "higher" functions) was reduced to the lowest possible common denominator. This strategy is called "reductionism," because it reduces the higher to the lower and the complex to the simple. It is also a radical reversal of the real order of nature, putting everything upside down on its head, and maintaining that there is no essential difference between anything anymore, that everything is lifeless, colorless, inert and mechanical "matter." Scientists were satisfied that they could prove that an organism is "nothing but" a more complicated system whose principle laws and operations are exactly the same as (what they imagined) happens in the most simple and mechanical physical systems. What is happening now in science is really a *reversal of this reversal*: instead of making everything lifeless, inert and "dead," everything turns out to be more akin to life and *becoming*.

Note well that also here the gap between life and lifeless matter is closed, but the other way around: instead of a down spiraling reduction, we see an upward swinging movement that opens everything to life, change, growth, becoming and evolution. The science of matter, life, and the human demonstrates that "life processes" take place not only in the limits of the cell's membranes, but in metals and crystals, elementary particles and galaxies, world markets and dynamic social systems. This is at the same time also a reversal of the old vitalist beliefs. The last vitalists believed life to be a hidden, mysterious fluid that operated only within individual living organisms, but in contemporary science, the wonder of life seems *spread throughout* the entire universe. Properties and functions believed to be restricted to living organisms alone are now discovered in all other systems. It seems that vast fields of "non organic life" are penetrating deeply into realms that were traditionally guarded as the realms of lifeless matter. Self-replication and reproduction, spontaneous self-organization, regeneration and formation are becoming the rule and not the exception in the open systems of physics and chemistry.

Moreover, we are increasingly realizing that it is possible that the **whole universe may be an open system** in a state of continuous becoming. It is also becoming clear that universal entropy (and gravity, its twin sister in classical physics) play not only a subordinate role in the laws of this becoming, but may be understood at last to be integral parts in an open universe. In open systems that constantly receive and exchange energy, information, and formative forces from their environments, entropy is just a *register* that measures the amount of disorder that a system produces as a byproduct of its self-generation and development. That is, it tells us *how much* disorder is needed in order to create new order out of fresh chaos. It is therefore not totally unthinkable that the whole universe is just such a "non organic living organism," and that this is the reason why "matter" everywhere shows signs of life. After all, this matter has been deposited and actualized in and through the forces of the universe. In other words, organic life as we know it on earth may be a *local embodiment and actualization* of universal becoming. Life is embodied materially in its fullness in the living organism, but really expresses itself everywhere.

Then organic life will not be seen any longer as a wild and meaningless violation of the eternal rules of a lifeless, death-full, abstract "matter," but rather one local expression of universal forces everywhere at work in the cosmos. This life is present in all material systems, but expressed in different ways that are not fully, nor necessarily, organic. Organic life is life's embodied and individualized form, located within membranes and skins, and operating in its environment with self-producing "inner" forces. Yet what is inwardly active *in* the organism is spread all around it as well, not only in its organic environment, but also through the whole earth and cosmos. **What if life, intelligence and consciousness are not only**

embodied in material-organic forms, according to the strategy of individuation, as we know them on earth? What if they operate everywhere as forces and potentials of becoming that stream freely through the whole universe and express themselves in all forms and states of matter, in all forms and states of cosmic evolution? If this turns out to be the case, then the physical world is going to be perceived in the near future, by the physical sciences themselves, as a world suffused and filled with streams of non-organic life. And if indeed in this world humans begin to feel a little more "at home," it will be a home not closed upon itself as an island in a foreign and hostile cosmos, but an open and light filled home, fully integrated with the real forces of the universe, flowing in and out through its transparent and intelligent membranes, with the tides and ebbs of earthly and cosmic becoming.

Philosophical Embarrassments: Is the 'I' Within or Without?

Such insights bring with them extensive changes in our understanding of the mutual relationships, adaptations, and information exchange between the organism and its environment. Great difficulties for philosophy, psychology, and the behavioral and social sciences, were caused because of the essentially dualistic theory of representation. According to this theory, any organism, including the human, is said to represent the 'external world' in its 'inwardness.' Life, cognition, consciousness, and intelligence as a whole, were understood as something taking place within the limits of membranes, skins, or brains, possessed by a being essentially separated from its environment, and from within these limits, reactions could issue forth to answer external stimuli.

This questionable idea has also produced inadequate accounts of knowledge acquisition. This conception believes that a true proposition is one that a subject creates inside his mind or brain of an external or internal object or process. It is thought to be "inside" a body or brain, and yet, it must be able to represent, exactly, an "external" fact or object. But this subject is also *self*-conscious: it makes representations of itself as well as of external objects. Now this subject would have to be a rather miraculous being, like a snake that bites its own tail, because it would have to be able to view itself from the outside to snap a mental picture of itself, in the same way it pictures the tree. This theory presupposes, therefore, a self-representing subject, that, besides making mental pictures of external objects, makes also internal mental pictures of itself. But if it is a real self that creates all those representations, what it is made of? It cannot also be made of a mental picturing stuff. And where exactly is it located physically?

This theory must therefore assume that the information from the external world somehow infiltrates me through my bodily-spatial boundaries in which I am supposedly encased, through my skin and my senses. Then it streams to a central control room, the brain, where an inner observer synthesizes a single, coherent, and exact mental picture of an external object from a mass of very different impressions and information. The representation will be 'true' and the propositions based upon it will share this truth, to the extent that it can be proven that such an exact replica has been created.

Now one major difficulty with this theory is that all the items taken in through sensory inputs have nothing in common with the object as represented in my real life experience and observation. The physiological research of the neurological processes in the senses, nerves, and brain discover infinite molecular, chemical, and electrical events on the way from the 'external object' to the 'inner subject,' but none of them has the slightest similarity to either the object as represented, or the supposed subject that represents them. For example, the red I perceive is not at all similar to the electrochemical processes in the eye and brain; and the height of absurdity is apparent when I ask *what color* the self-representing subject could be and *how* its color could possibly be sensed. There is no hint at all of the object itself in any of the molecular and electrical processes in the senses and brain, nor do we have any subjective experience of a little man that is uniting all of them into one picture. For us, the picture of a tree is a given experience, and we are not conscious at all of the various sense and nerve processes that take place in our bodies and brains while we observe the real tree. The brain functions as a self-organizing, self-unifying machine, and has absolutely no need of such an inner subject. And of course all the various findings of neurological science are also objects of our real experience, no less than the tree out there.

When the scientist studies the processes in the brain that take place when a person observes the tree, she is consciously observing her sense perceptions of the nerves and brain processes she studies. While she studies the brain processes before her, perceiving them and thinking about them, in her brain very similar brain processes are naturally occurring. So what is "outside" for her and what is "inside?" Is there any meaning at all to these spatial and dualistic concepts and representations? When I observe my own neurological activity in real time through a brain monitor, or for that matter, if I study the microscopic structures of someone else's brain or a piece of wood, I never encounter an inner self. I never encounter an inner observer and catch it in the act of representing.

The real experience is indeed very different from what is imagined and theorized above. Even though I imagine that I am the owner of 'my' brain and possess it like I own this chair, it is apparent that **in** the moment of experience, without reflecting on 'my own' activity, I wholly *forget myself*. Actual experience and reflection about it are two distinct experiences. *While* sensing and thinking I experience myself wholly given to the processes I experience, touch, and think about. Phenomenologically speaking, without preconceived ideas of an inner self, **I am wholly inside the outside,** when I see, hear, and touch but also think, feel and will. And conversely, the red I see and the tune I hear are wholly within me. But actually, all spatial metaphors are of no use here because as long as I am in the experience I lose any sense of space (and also of time). I represent, in other words, only a holistic, non dualistic, experience after it is finished. But again, this reflection is in itself a new, real, non-representable experience. In the moment of *any* experience, I am again one with my object, or the object is again part of me.

In other words, the theory of representation says that I sit comfortably inside myself, observing the external world from a distance, gathering information about it, and composing a picture from it that fits the 'original object.' However, my real experience is something *totally different*. Such a represented subjective 'inner self' is precisely what I **do not** experience when I plunge into the real processes I experience, observe, think, or feel; and only when I reflect back on my *past* experiences I can make a representation of them. But in the process of making a representation of the past, I am actually, in living time, again plunging into and becoming part of the processes which I observe, sense, think and so on. Nowhere do I find this separate, self-representing little person, located inside my body, brain, soul, psyche or mind, nor do I ever see or find the site from which I as an inner being really look outside myself and make a picture of the world as I do with a camera. On the contrary, I find out that when 'I' feel most strongly alive and real is not when I make a mental picture of myself or of you, but when I become red or blue, me or you, the tree or a neural pathway observed under the microscope. The whole point is that when I am actively sensing, feeling, observing or thinking in *real time*, I am as if half asleep, in so far as my reflective self-consciousness is concerned. I am fully conscious of my experiences, of course, but I am not self-conscious, that is, I am not able at the same time to experience and reflect on the **same** experience. The reflecting itself is always a new act and experience in itself, which cannot be, in the same time, reflected upon. I am simply one with the world whose content I experience, and this is the reason why I forget, indeed, not my *real* self that is one with the world, but my self-conscious representation of myself and of the object while this happens. A representation has its justified place in the cognitive process as a whole, so long as we do not forget that we make it only *after the real experience is over*. We can truly speak about a mental picture, a representation, either of an external or internal experience, only if we are aware of the fact that we create it when we reflect back on an experience that is finished and that we are now out of the real activity in which we were immersed as long as it lasted. Furthermore, I would have never naturally come to the conclusion that now, when I make myself a conscious mental picture about the past event, that in this representation I have the real thing or the real me. On the contrary, I would have said: I was part of real being, I was part of the sum, but this happened precisely *before* I began to reflect about it, and make a representation out of it. When I was *really thinking*, or *perceiving* real red or blue, I was not creating mental pictures because I was *inside the outside world*. I build mental pictures and representations about it after the fact, and only then I can say that I have come out of real life and landed on the secure and subjective shores of representation. But I will never say that these representations are the real I or object, and I would have never come to the conclusion that "I reflect, therefore I am," but would have certainly said: I reflect, therefore I am not (real, but only a mere picture of the real).

So, then, finally, what if this inner isolated subject is simply not real, only imagined, what if it is not there at all? What if the whole conception, the whole question, was wrongly put from the beginning? At the very least we have to admit that the whole theory of the subject and of representation breaks down, because its central assumption is precisely something that is never found, on any level of my perception, thinking, cognition, or judgment. The cognitive realities that the theory describes do not really exist. Only a wholly new approach to cognition, that only begins to emerge today, may attach for the first time some real meaning to such terms as *picturing, reflecting, representing, coding* and *decoding*, but what is certain is that in accordance with the older theory of representation they lack any real content and meaning.

In short: what the theory says I am doing, I never do, and what I am really doing is something wholly different, something that is precisely *unrepresentable* in the way the theory of representation would have me imagine. This is the reason why the philosophical, psychological, cognitive and neurological investigations that in the last centuries sought to find a real cognitive process and a real image of what the human is and how it becomes, have ended in such embarrassments. Nowhere could such a represented-representing "subject" be found, nor a representation of the external object in this subject or in a brain, and nor, finally, any representational process that would even begin to be similar to the things we really see, hear, touch, experience and cognize.

Actually, David Hume already discovered that Rene Descartes' "I think" subject could never be found at all in his mental and soul researches. Hume's conclusion was that there is no mentally substantial "inner self" that can be grasped, taken hold of, and represented in inner consciousness as an object. And would it not have been for Immanuel Kant's cunning recovery of this subject, through his transcendental analysis, a direct phenomenological line *would have led us* from Hume to Nietzsche, Steiner, Husserl, Bergson, Heidegger, and on to the 20th century poststructuralist thinking that will be studied in our later chapters.

In chapter 3 we will provide an updated answer to Hume's *I think therefore I am not*: namely, that if an 'I' cannot be represented in inner experience, closed in its own skin, membranes, or brain, as an isolated subject, this must not at all mean that there is no **real** 'I,' real world, and real connection between them, only that they cannot be subjects and objects of our representational cognitive powers. **If the 'I' cannot be represented, this is not because there is no 'I,' but because the 'I' belongs to the realm of the real and not to the realm of representation.** Such realism, however, is the same unbiased realism that is now establishing itself in all fields of science and thinking, and it intends to find the real not where it is (not) represented, but to find the real where it really is, namely, everywhere *flowing* and *becoming* in the outside of the within, and the inside of the outside.

Embodied Cognition: The Santiago School

An important pioneer of the new scientific revolution in the field of cognitive studies was Gregory Bateson, who in *Steps to an Ecology of Mind* (1972), dared to advance a "cybernetic" vision of intelligence and thought. He suggested that cognition is found not only in the mind but in all natural processes in the external world, and that, more importantly, intelligence is immanent not only to organic life, but to all physical systems. This concept has increasingly become prevalent in biology and cognitive science. At the same time, two Chilean scientists, Humberto Maturana and Francisco Varela began to publish their common work about *autopoiesis*, and the so-called "Santiago" school was founded, whose influence cannot be exaggerated. And while some aspects of their early version of autopoiesis (self organization) could be said to suffer from some residues of the traditional representational conception of organic life, the more mature version amounts to the first systematic and substantial effort to equate mind and cognition with life as such. Their famous equation says it all: "*Living systems are cognitive systems, and living as a process is a process of cognition. This statement is valid for all organisms, with or without a nervous system.*" (*Autopoiesis and Cognition*, 1980) As we shall presently see, such conceptions are expanding today to include the human being, the body, the mind and consciousness, as part of the same, radically new, worldview.

Maturana and Varela maintain that "cognition is embodied," or that it can be found just where it is active. It is synonymous with all life processes, and mind is not restricted to higher organisms, endowed with nervous systems or even a minimum requirement of a rudimentary sense apparatus. On the contrary, any life process is intelligent, in the sense of nutritive, metabolic, respiratory, reproductive and adaptive strategies and cooperation. No doubt many people are still spellbound by the Cartesian idea of a mind concentrated in a tiny point inside of their heads, while the rest of the universe is only dead, unintelligent matter. But even if we dismiss that bias, we may still find it hard to accept that all life thinks and that all thinking is alive. The problem is that we cannot suppose that nature thinks as we do (and to be honest we really have no clue *how* we think), but rather that it thinks in its own way, just as it weaves, creates and evolves.

Now autopoiesis first tried to defend the organic life of any organism against purely mechanical biological thinking, which was a common part of the old reductionist paradigm outlined above. This was helpful. The inherent capacity of the organism to self-regulate its life and its interrelationships with its environment, in an active, even creative way, had to be defended and deepened first. But indeed, from the moment Maturana and Varela crystallized the formula that living systems are cognitive systems, they had begun to widen their concept of autopoiesis far beyond the organism's membranes. The whole project now shifted its emphasize from defending the rights of an isolated organism to an independent and autonomous existence, on to a vision of an **organism + environment + cosmos**, as a new cooperating, intelligent, and emergent system. Many joined them in this work which continues today. This new concept of life suggests that the organism is no longer an island in a hostile environment, struggling to survive, but that it is a droplet, stream, or better yet, a network of flows and currents, making up the very life of the ocean in which it is submerged. Furthermore, this life of the organism in the ocean of life is "meaningful," or "cognitive," in a fundamental biological and metabolic sense. Varela put it this way in the article *Organism: a Meshwork of Selfless Selves*: "There is no food significance in sucrose except when a bacterium swims up-gradient." He says that this "surplus of signification" is "enacted" externally as the bacterium swims towards the meaningful sense of the sucrose. In other words, this means that living cognition is not an inner representation of an outer fact, or sign, but an active enactment, in-volvement and engagement in potentially meaningful living systems.

In Evan Thompson's *Mind in Life*, we will find very similar conceptions. For Thompson, autopoiesis, cognition and making sense, in sensory-motor enactment, has the following four main characteristics. First, the organism is primarily and essentially open to its environment, not closed in a membrane or skin that must somehow struggle to find ways to "communicate" with the external world, overcoming its isolation. Second, the organism's involvement and integration in its environment is meaningful, and sense creating, as well as regulated according to its inner metabolic needs. Third, the movement of the organism in space and time is a result of the above, which means that the real life of a real organism, is embedded and enacted in a life world, in which the outside and the inside correspond to each other and serve each other because both are sharing the same essential cognitive/ living characteristics. Fourth, autopoiesis is also necessarily "self-domesticating" and juvenilizing (neotenizing) because it regulates self-environment relations in favor of the organism's developmental and evolutionary needs for its adaptation and survival, which means, in the new emergent evo-devo synthesis (to be discussed below), that it must "simplify in order to complexify," or "involve in order to evolve," and vice versa.

Now, it may be objected, that even such positive terms as *adaptation, symbiosis, cooperation, synergy* and *enactment*, still carry with them residues of dualistic body/organism versus mind/environment conceptions, and therefore also still harbor some after-effects of classical reductionism. But we are in the midst of a process of change and cannot expect everything to happen at once. A whole new language is now emerging, and it is still in the making and, probably, will always remain so, as our understanding of existence and becoming, now freed from the fetters of reductionism, only begins to awaken to the remarkable horizons that open before us. This is a vast and radical synthesis in the making, converging at the same time from all directions, like an intellectual "super storm," from AI and robotics, genetics, the newest researches in neurological and brain sciences, populations and ecology, ethology, sociology, ethnology, psychology, cognitive sciences, poststructural and postmodern philosophy as well as the study of altered states of consciousness. If I had to characterize this new synthesis in one formula, I would extend Maturana's and Varela's formula, as follows, "*All things and systems (material, living, and conscious) are cognitive systems, and all material processes are cognitive processes. And this statement is valid for all things and systems, with or without material embodiment.*"

Reviewing further evidence from present thinking and research, this definition may serve as a more updated starting point for a whole new conception of *becoming*, on all levels of our existence. This was beautifully expressed in the conclusion of their book, *The Tree of Knowledge*: "*We have only the world that we can bring forth with others, and only love helps bring it forth.*"

Out of Our Heads

Another pioneer of the second scientific revolution is J. J. Gibson. He represents a significant junction in 20th century evolution and transformation of science because he was a student of the Gestalt psychologist Kurt Koffka, and has been influenced by Lewin, Piaget and Boring.

In his *The Ecological Approach to Visual Perception,* (1979) he offered a radically new, and timely, conception of perception and perceptual cognition. Let us listen to his words, and notice the refreshing change of point of view breathing through them. With Gibson the new paradigm has reached a stage in its becoming that allows it to enter boldly into traditional strongholds of reductionist, dualistic science, most specifically, into the fields of the study of perception and cognition. With Gibson as our guide, we are definitely and irrevocably exiting our representational cage:

> I have described...what the environment affords animals, mentioning the terrain, shelters, water, fire, objects, tools, other animals, and human displays. How do we go from surfaces to affordances? And if there is information in light for the perception of surfaces, is there information for the perception of what they afford? Perhaps the composition and layout of surfaces constitute what they afford. If so, to perceive them is to perceive what they afford. This is a radical hypothesis, for it implies that the "values" and "meanings" of things in the environment can be directly perceived. Moreover, it would explain the sense in which values and meanings are external to the perceiver. (p. 127)

Gibson maintained the active character of the perceiver and expanded his vision theory into a cognitive theory of the type advocated by Piaget's developmental psychology, Gestalt psychology and phenomenological studies of perception that found such a productive expression in Merleau-Ponty's masterpiece, *The Phenomenology of Perception*. Gibson remarks:

> Perceptual development and perceptual learning are seen as a process of distinguishing the features of a rich input, not of enriching the data of a bare and meaningless input... Above all, the puzzle of meaning and value in perception takes an entirely new form. If what things afford is specified in the light, sound, and odor around them, and does not consist of the subjective memories of what they have afforded in the past, then the learning of a new meaning is an education of attention rather than an accrual of association.
> (*The Senses Considered as Perceptual Systems*, 1964, p. 320)

Also, the following quote from the philosopher, Alva Noë, can serve as a relevant and symptomatic expression of some of the currently developing conceptions of life, cognition, and consciousness that emerge from biology:

> We ourselves are distributed, dynamically spread-out, world-involving beings. We are not world representers. We have no need for that idea. To put the point in a provocative way, we are, in Merleau-Ponty's memorable phrase, "empty heads turned toward the world." And as a result of this, our worlds are not confined to what is inside us, memorized, represented. Much more is present to us than is immediately present. We live in extended worlds where much is present virtually, thanks to our skills and to technology.
>
> (*Out of Our Heads: Why You are Not Your Brain, and Other Lessons from the Biology of Consciousness,* Alva Noë, 2009*;* see also his previous study*: Action in Perception, 2004*)

What is so remarkable about the present "event" in the sciences, is that it demonstrates everywhere the fact that this event is not a mere change, not even a revolution, but a real "reversal" or "inversion" of the previously accepted views of the world and the human being. Such reversals of Descartes' *cogito*, expressed by Merleau-Ponty's careful and exact phenomenological investigations, are now being rediscovered in so many present empirical scientific studies. Indeed, as we shall see in the coming chapters, this peculiar expression "empty heads turned towards the world" has much in common with the event not only in science, but also in philosophy, art and history.

One important field of research and application, without which many of the present physical and biological insights would not have come about, is the field encompassing cybernetics, AI, and robotics. Its contribution to the emerging new conception is fundamental, as are its own struggles to emerge from and overcome its own mechanistic prejudices. Together with ecology, biology, and ethology, it helps us to release our thinking from the mechanical concepts of intelligence, mind, communication and information. It was common to view the brain and cognition according to analogies we drew with the technology we recently created. So it was natural to compare our mind to a computer, since the brain seems to be the best we know so far! But this tendency to measure nature according to our latest technological achievements, in order to understand nature and human nature is as understandably anthropomorphic as it is uncreative and eventually futile. Nature has been working already some billions of years, and it is really childish to expect that our own inventions, achieved in recent evolutionary seconds, will serve us as models to understand nature. But today we are more humbly inclined to reverse this attitude, and admit that we are nature's pupils, though we may be very proud of our own achievements.

Robotics is one field that really brings this lesson home to us, as there we are trying to *implement* our theoretical concepts of brain, behavior, mind and cognition, and because in this field there is much less room for entertaining false conceptions. After all, in building robots it becomes practically very clear, what we can and cannot do, and what we really understand concerning the most basic functions of cognition. If we survey the latest developments in this field, we can learn a great deal in this respect (For example, see *Sharkey and Ziemke, 2000; E. Hutchins, 1995; R. Brooks,*

1991; M. Minsky, 1987; A. Clark, 1997, and M. Wheeler, 2005).

Andy Clark is a leading cognitive scientist whose interpretations of the present developments are very timely. He came to the conclusion that we must place the organism in its environment in a **mutually** holistic way. This means that the "internal environment" must be defined by the "external environment," but also vice versa, because the organism is actively changing and forming its world, according to its needs, and it must be understood as an "active adapter" or "adaptive responder." Here *body-brain-world* is seen already as one complex, holistic, mutual, intelligent and co-evolutionary system, taking a long stride in overcoming mind vs. world, mind vs. matter, and life vs. cognition dualisms. (See also Andy Clark, "Re-Inventing Ourselves: The Plasticity of Embodiment, Sensing, and Mind" in *Journal of Medicine and Philosophy*, Volume 32, Issue 3, May 2007, pages 263—282. Also see his latest book: *Supersizing the Mind: Embodiment, Action, and Cognitive Extension*, 2008 for the latest updates and enlargements of his "extended mind" concept).

In 1991, Rodney Brooks published his paper *Intelligence without Representation,* strongly challenging AI, Robotics and cognitive science to look deeper into accepted mechanist notions in these fields. Brooks, the head of the AI lab and the Hominid Robotic research group in MIT, has also described some of his practical learning experiences in this field. He tells us that in the first years of AI, the scientists were feeding the computers with complex programming that simulated solutions to complex problems. Some results were impressive indeed, for example, the development of IBM's *Deep Blue* that defeated Kasparov in 1997. But

Brooks says that, to the surprise of the researchers, they began to recognize that chess playing is a far more simple activity than, for example, (and here comes the humbling part) making soup. "Robo chess needs only logic and information, but Robo chef without a spark of creativity, intuition, and an aesthetic sense, will be nothing but an expensive food scrambler with a programming option." Brooks claims that our intellectual, logical and analytical problem-solving capacities, our greatest pride, are much easier to imitate and implement with robots, than the most basic perceptual and intuitive actions of any little child.

Just consider this fact: scientists and developers of AI and robotics must turn their attention to the study of the most basic development of babies and little children, before they can even start to *imagine*, let alone implement, the simplest robotic imitation of human behavior! Some of them believe today that in order for robots to ever acquire human like intelligence, they may have to undergo a development similar to that of babies. But as we know, children learn though imitation of adults, brothers and sisters, and through trial and error. For this you need social interactions. Cynthia Breazeal, a robotic expert at MIT, says that this is the reason why her *Kismet*, the famous robot developed through her work, is programmed to search for sensory inputs—sounds, movements, colors—which it calls by means of baby like, inviting, facial and vocal gestures. If it succeeds, and a passing human comes to play with it, its "social drive" is satisfied. If not, its level of activity drops and it is programmed to start a new hunt for learning adventures and for initiating new social intercourse. "The point is," says Breazeal, "that Kismet is trying to engage his partners in such a way that will advance its ability to learn and develop its skills... this helps the robot to learn the social meaning of its actions. The goal is that Kismet will learn not only to 'think' for itself, but will also understand that its actions have implications—as any child must learn."

"It's not what's in the Brain that counts, but what the Brain is in."

This statement is taken from Michael Wheeler's book, *Reconstructing the Cognitive World: the Next Step* (2005), in which he also reports on the findings of an MIT research group concerning the shifting paradigm in AI, and the cognitive and neurological sciences. Using Heidegger's phenomenology as philosophical resource, they are striving to formulate a new and consistent conception of non-Cartesian cognition. The new cognition would have four complementary characteristics. It would be "embodied and embedded, enactive, extended, and affective."

With the term "embodied" they wish to break from current computer analogies, in which intelligence is seen as a central processing unit (CPU). In this highly centralized and mechanical model, the brain and human cognition function like a computer, carrying out the instructions of an already given programming. The CPU carries out each instruction of the program in sequence as programmed to do, and performs the basic arithmetical, logical, and input/output operations of the system. But Wheeler and his co-workers are employing more advanced modeling, called "dynamical systems theory," that has the advantage of being able to describe distributed, de-centered, neurological networks and processes. In this way they try to envisage a "self" that is not locked into a centralized, brain-bound nervous system. Thus, "embodied" cognition becomes "embedded" cognition, as Bateson would have it, not representing-represented, but active and actual wherever and whenever an exchange and processing of intelligence, information and knowledge are at work. In conventional thinking, cognition is imagined to be located like the middle of a sandwich, crushed between sensory input and motor output. The subject would then be the little man we spoke of earlier, taking in the input from the senses, processing it, forming representations that exactly reflect it, and transmitting them to the motor centers. After this, the little man would form motor representations whose purpose is to fit and serve the motor reactions that should reply to the input according to the adequate representations formed.

In contrast, when intelligence and cognition are embodied-embedded, they are also "enactive," because they are distributed differentially along the functional pathways and spectrum of the whole field, being involved and engaged, simultaneously, in all cognitive aspects, without any need to hypothesize a centralized, preprogrammed, mechanical center of computation, and without a self-referential, representing "inner subject."

When the complex and dynamic system(s) composed of *body-brain-mind-world* is woven dynamically in such a manner, cognition is also woven, and becomes a world-wide-web, (or as Clark has described it) a truly "extended" mind. The concept of a ghostly disembodied self, an abstract mind or a Cartesian thinking substance disappears from this world entirely, along with its twin: matter without life or cognition. If cognition is biological (but note that this limitation will eventually disappear), and life is cognitive, then being *affective* indicates that meanings are objective world processes. The human world is not reduced but instead the world becomes more human. The human is no longer an isolated and absolutely unique creation of reason, but now shares a valuable universe with all.

Some decry this as an anti-humanist trend, because they are focused one-sidedly on what we are losing, when our absolute uniqueness and mastery of nature disappears. But as a matter of fact, our gain is twofold: first, losing this anthropocentric position is a blessing not only on a cognitive level, but on a moral one as well, because it releases the human hermit-subject from itself and frees its inner forces to join its natural and social-human environment in a wholly new way. Second, why should we complain about the loss of the old human, if we are now beginning to find that all the good things of life, that we believed to be solely in the possession of mankind, concentrated inside the human mind and heart, are really spread out and distributed in nature and in the cosmos as a whole?

Part Three: The Involution of the Theory of Evolution

IN THE CLASSICAL MECHANISTIC GENETIC PARADIGM, the genome was imagined to be Cartesian hermit, located in its isolated inner cellular space, a place perfectly fit for analytic experimentation and intellectual quantification and formalization. Now it is important to remember that this approach to biology had its time of fantastic progress and achievement (recall Ben-Jacob's remarks about the paradox of the creativity of those denying creativity), because reductionism as a whole was perhaps, despite some of its ideological commitments, an epoch of exemplary creative scientific progress. Today, however, scientists clearly realize the limitation of both the theoretical and practical reductionist and mechanist approach to genetics, development, and evolution. Therefore they are beginning to study the interrelationships between genome and cell, organism and nature, and nature and universe; all of which are increasingly seen as an amazingly complex emergent and innovative process of becoming. And this conception of cosmos, nature, and ecology has influenced our view of the genome, and suggests a very different notion about how DNA, development, and evolution actually work together.

As Richard Lewontin puts it in *The Triple Helix: Gene, Organism, and Environment (2002):*

> Individual development is not an unfolding, and evolution is not a solution to present problems. Rather, genes, organisms, and environments are in reciprocal interaction with each other in a way that **each is both**

> **cause and effect** in a quite complex, although perfectly analyzable, way. (p. 61)

From an atomic, analytic, and linear concept of genetics, we are moving to an intricate and multileveled notion in which the genome is not so much the source, but more the means or instrument of evolutionary variation and change. The genome is not the originator of what it codes, but is part of a much more universal, still largely unknown, coding process of becoming, woven over billions of years, as a web of both organic and non-organic matter, life, and conscious mind. It is a tapestry full of information, order, and beauty as well as surprising and unpredictable spontaneity. It appears now that *genome-organism-environment* is essentially a holistic and integrated complex system in which genetic change is not directed in a single, linear way, from DNA to organism, but also the other way round. Not only does the cell's own inner environment determine the DNA, but also its external environment.

This implies multi-directional feedback relations at all levels of the system, ranging from DNA, RNA, proteins, cells, and networks of cells, organs, *and* environment. The environment can actualize and activate DNA in a variety of ways producing very different outcomes. This means that a process of mutual influence takes place, allowing genuine innovation, and creative emergence. What is essential in this process is that the environment is not the enemy of the organism, and so autopoiesis should not be understood as an organism's way of harnessing an alien environment to serve its own needs.

Variation and potentiality are found in the 'organism,' in the 'environment,' as well as in the mutual relations between them. This means that the *actualization* of existing genetic information is part of an extended capacity and plasticity, in which *existing*, but *not expressed* genetic variations, may be actualized, but also *new*, innovative mutations. In addition, the interplay between past and future becomes a *rich new dimension*, in which the presently perceived or actualized phenotype is but one expression of a whole 'host' or 'swarm' of possible variations.

Brian Goodwin describes this promising conception of genetics as follows:

> What counts in the production of spatial patterns is not the nature of the molecules and other components involved, such as cells, but the way these interact with one another in time (their kinetics) and in space (their relational order — how the state of one region depends on the state of neighboring regions). These two properties together define a field, the behavior of a dynamic system that is extended in space—which describes most real systems. This is why fields are so fundamental in physics. But a new dimension to fields is emerging from the study of chemical systems such as the Beloussov-Zhabotinsky reaction and the similarity of its spatial patterns to those of living systems. This is the emphasis on self-organization, the capacity of these fields to generate patterns spontaneously without any specific instructions telling them what to do, as in a genetic program. These systems produce something out of nothing. (*How the Leopard Changed its Spots*, 2001. P. 51)

One possible way to interpret this new concept of "morphogenetic fields" is to say that we discover a new genomic double structure, but in this case it is not about two strings of DNA but of two levels of genomic as well as possibly also DNA functions, one actual and the other potential. One stores and codes accumulated, actualized past evolutionary change and variation, while the other keeps open a formative potentiality that will receive the next mutation or adaptation. This potential genomic function, operating along with selective pressure allows the creative emergence of an entirely unique evolutionary invention. This also means that the genome can be seen as a non-local and non-temporal universal pool, not only of stored evolutionary information, but of *evolutionary potential*, spread out in the open world in space and time. In space, it is seen as part of an extended and dynamic genomic network, which includes all the genetic material on planet earth as one infinite pool of information and potential for originality, shared by all living organisms. In time, DNA is coding not only actualized genetic information, but a great deal of conserved, even primitive, genetic potentiality, shared across phyla throughout historical evolution. (We will come back to this reserved potentiality in time, when we study the key role of involution, paedomorphosis, and neoteny in evolution in what follows).

Among many other factors that have contributed to this new concept are investigations of Bacteria over the last two decades. Let us look at them more closely. Roger Lewin reports:

> One of the experiments involves taking colonies of *E. coli* that are incapable of metabolizing lactose and exposing them to the sugar. If the lactose-utilizing mutants simply arise spontaneously in the population and are then favored by prevailing conditions, then this would lead to one pattern of new colony growth. A distinctly different pattern is produced if, under the new conditions, the rate of production of lactose-utilizing mutants is enhanced. The observation is something of a mixture of patterns, indicating that directed mutation appears to be occurring. *'This experiment suggests that populations of bacteria...have some way of producing (or*

selectively retaining) only the most appropriate mutations,' note Cairns and his colleagues.

(Roger Lewin, *"A Heresy in Evolutionary Biology,"* *Science*, 241: p. 1431, 1988)

Research with *E. coli* at other labs is producing similar heresies. (See Hendricks, M. *"Experiments Challenge Genetic Theory," Science News,* 134: p.166, 1988. Also: Cherfas, Jeremy; *"Bacteria Take the Chance out of Evolution," New Scientist,* p. 34, September 22, 1988.)

Eshel Ben-Jacob's work with bacteria colonies has led him to introduce "paradigm changing" ideas into the present developmental-evolutionary debate. It is worthwhile to quote from his words at some length here:

> It is known that in a stressed colony, some of the bacteria become competent by rendering their membrane more permeable to genetic material, while other bacteria go through lysis: break open and deposit their genetic material in the media. In addition, direct genetic connections between the bacteria are formed by means of conjugation or transduction. We propose that these features indicate that the stressed colony turns into a genetic network, which is the highest level of colony cooperation. To emphasize that the network is composed of agents (each genome is by itself a cybernetic agent), I refer to it as a "genomic web." I further assume, that in order to establish the genomic web, the bacteria produce (or activate) special cybernators enhancing the efficient and sophisticated genomic communication. Once formed, the genomic web is a "super-mind" relative to the individual genome. Thus, a paradox for the genome is a solvable problem for the web. The web, being more complex than the individual genome, can design and construct a new and more advanced genome relative to the original ones, i.e.

> perform a vertical genomic leap. Such a leap is best described as a cooperative self-improvement or cooperative evolution.

The concept of a "genomic web" is central to the new paradigm, as an example of the new insights into *becoming as such*. If a cooperative genome means that inherited evolutionary "wisdom," or intelligence is shared across chromosomes, nucleotides, membranes, indeed, why not across phyla and species, space and time? What if all genetic information is, in principle at least, able to be shared across all organic limits and barriers? Yet, if this were so, it would not be primarily through physical outlets, but rather because (as we shall indicate below) all matter, life, and intelligence shares the same basic universal virtuality or open-world wholeness.

Ben-Jacob believes that this new picture of genetics suggests a new vision of evolution itself:

> I hope I was successful in convincing the reader that Vitalism is not the only alternative to Darwinism. I propose a new option: that of cooperative evolution based on the formation of creative webs. The emergence of the new picture involves a shift from the pure reductionist point of view to a rational holistic one, in which creativity is well within the realm of Natural sciences.
>
> (Ben-Jacob, *Bacterial wisdom, Gödel's theorem, and creative genomic webs*, 1997, and see also: *Bacterial self-organization: co-enhancement of complexification and adaptability in a dynamic environment, 2003).*

It can be noted that Ben-Jacob's view seems to be in agreement with Kauffman's "order for free," which suggests, "In sufficiently complex systems, selection cannot avoid the order exhibited by most members of the ensemble. Therefore, such order is present not *because of selection* but *despite* it."

Evo-Devo and the Next Step

Already in the 1970s, mainstream thinkers of evolution, like Stephen Jay Gould and Niles Eldredge, maintained that evolutionary biology would come to rely increasingly upon changes in developmental genes (that is, genes open to, and directly responsible for, the "bi-directionality" or mutual exchange between genome, organism and environment, mentioned above):

> We do not see how point mutations in structural genes can lead, even by gradual accumulations, to new morphological designs. Regulatory changes in the timing of complex ontogenetic programs seem far more promising—and potentially rapid, in conformity with our punctuational predilections. The near identity of humans and chimps for structural genes, and the evidence of major regulatory change evidenced by human neoteny provide an important confirmation. (Gould, S. J., & Eldredge, N., *"Punctuated equilibria: the tempo and mode of evolution reconsidered,"* Paleobiology 6(1), (1977), pp. 115-151)

Notice that human *neoteny* (holding on to typically juvenile features into adulthood by slowing down development) is invoked here to exemplify a general evolutionary trend, because the **structural** genome of human and chimpanzees is for all genetic purposes identical, yet their development **in time** is regulated in a remarkably different manner, **retarding** it in comparison to the chimpanzee. Humans seem to slow down in displaying central aspects of bodily development, which contributes to bipedalism, uprightness, speech, brain growth and so on, and as a result of these, also a postponing of the timing of sexual maturity.

The same point has been made recently by James Shapiro, expressing the conviction that the scientific and theoretical foundations of genetic-evolution and development are undergoing profound change.

> The point of this discussion is that our current knowledge of genetic change is fundamentally at variance with neo-Darwinist postulates. We have progressed from the Constant Genome, subject only to random, localized changes at a more or less constant mutation rate, to the Fluid Genome, subject to episodic, massive and non-random reorganizations capable of producing new functional architectures. Inevitably, such a profound advance in awareness of genetic capabilities will dramatically alter our understanding of the evolutionary process.
>
> (James A. Shapiro, "A Third Way" at *www.bostonreview.net)*

Two other books capture the central meaning of the present revolution in evolutionary thinking: Evelyn Fox Keller's *Century of the Gene* (2000), and Mary Jane West-Eberhard's *Developmental Plasticity and Evolution* (Oxford, 2003). I will extract some of their essential elements and express them in a non-technical manner.

Trying to capture the meaning of the evo-devo as it pertains to the second scientific revolution, one can say that it is based on the discovery that important parts of the genome, in many and different organisms, are shared. That is, organisms spread over very long periods of time and across very different phyla, share similar, even identical, genomic patterns and information. What founded this conception was the discovery of the "homeotic genes." These genes are responsible for structuring development, by switching on and off processes of transcription that regulate gene expression. They are directly responsible for bodily plans, for example segmentation as well as spatial differentiation. They are very ancient genes, conserved without any significant change from times that antedated the division between the arthropods and mammals. An often-used example is the gene called "eyeless." If you transplant it from a rodent into an insect, it causes an induction of an eye, but the eye that forms will be an insect's eye. That is, the insect body determines that the eye formed will be an eye of an insect and not a rodent. This means that the genetic "chain of command" doesn't work only in a linear fashion, as was always imagined in classical genetic, from DNA to RNA and proteins, but also in the reverse order, that *transcription patterns* can regulate gene expression.

This discovery has led Evelyn Fox Keller, in her book, *Century of the Gene,* to write that we may now begin to consider the "mature mRNA transcript formed after editing and splicing to be the *true gene,*" and that this gene, and not the classical DNA hereditary gene, regulates development, acting back on the genome, and regulating from the cell's cytoplasm. In other words, it constitutes a reverse coding process: from the phenotype (cell's proteins) to the mRNA and to the DNA, instead of the customary, one-way mechanical conception of DNA-mRNA-proteins.

At this point we can consider West-Eberhard's main contribution, because she offered, to date, the most comprehensive synthesis of the present change, indeed, revolution, in modern evolutionary biology.

This change was already described above from various points of view. It is the revolutionary idea that *environmental influences on development, not mutation, are the first order cause of design*. West-Eberhard shows, by means of numerous examples, scanning all major fields of evolutionary and developmental research, which genetic changes occur when the environment induces a phenotypic change. This change works in such a way, that it directs and determines new selective strategies and implements them in pre-existing polygenic variations. Genes become "followers" and not the "leaders" of evolution! Then natural selection will favor the spread of a specific environmentally induced variant that has proved itself well adapted to the new environmental changes. This means that West-Eberhard's argument is that environmental induction is in fact more important than mutational chance accidents affecting the genes. West-Eberhard goes beyond the evo-devo because she expands the latter's focus, which is still based on genetic-molecular emphasis, and takes it one consequent step further. She asks, how have regulatory gene networks evolved? The organism as a whole is a many sided phenotype, networked in all directions in space and in time (to potential accumulated variations, not actualized but real nonetheless), and therefore its development is flexible and part of an open and wide world.

If it is so, and if diverse organisms share genes, how did their diversity arise? What is its real cause? The answer might well be found, West-Eberhard says, in different developmental networks that change the pattern of expression of the genes. But then, indeed, how do these developmental networks develop and evolve? This collectively shared genome network of networks "outside" the single DNA, cell and organism viewed as part of the natural and cosmic environment, doesn't function as a centralized, mechanistic "chain of command," or as a computational CPU functioning as linear actualization processes of already existing and fixed genetic information, but much more as inventive, or artistic processes. The scientific study of the cognitive processes of invention and creativity has more to tell us about development and evolution and involution of life, than we used to think. It appears that the unrealized, not yet actualized, potential evolutionary variations, changes and innovations, must have at their disposal a highly sensitive medium in order to express themselves. The genome must therefore be not only fantastically more complex in structure, function, timing and potentiality than was assumed thus far, but, furthermore, it must be capable of bidirectionality also in time, reversing evolution into involution, maturity into neoteny, that is, accomplishing perpetual juvenilizations as a major evolutionary strategy. At the very least, it must be able to do two things at the same time, that seem quite opposite to each other: conservation and codification of what is already actualized, in which inherited traits can be really preserved and transmitted to new generations, and, on the other hand, allowing for innovation and emergence, using potential possibilities open for future change.

The genome must be able to record, code, translate and actualize environmental changes, as an instrument played upon by the music of becoming, with an existing and stable enough keyboard embedded in a stable enough organized structure as a whole (the instrument), and yet letting the whole environment play a score that, as a matter of fact, is "written" when it is first played. What is more, it must be able to bring the two processes together and regulate and harmonize them at the same time: the environmental stream of possible change and the fixed hereditary stability. It has an openness for active, self-actualizing futurity and maintaining of past and presently conserved genetic structure and function. In short, it must at least operate the opposite structures and directions of actualization and potentialization, structuring and differentiating on one hand and at the same time de-actualizing and re-opening on the other hand to allow for new and unforeseen novelties. How can something be both structured and open? Can it be fixed and sensitive to new influences at the same time? Can it be actual and potential at the same time (and place)? A well formed, hardened instrument and keyboard and at the same time, also self-constructing and played upon by the music and winds of change from the nearest and farthest circumference and environment?

This is the view held by West-Eberhard in her review of Evolution in Four Dimensions, (by Eva Jablonka and Marion J. Lamb): *Dancing with DNA and flirting with the ghost of Lamarck*:

> But the revolution in evolutionary biology, if it is to occur, will concern not so much the mechanisms of heredity—in fact, one of the strongest points of the earlier Synthesis—but **the environmental causes of developmental variation of all sorts**, not just inherited variation; and it will focus on the relation of that variation to modern genetics, as research on gene expression and associated genomic processes (like methylation, chromosomal condensation, alternative splicing and environmentally modulated mutation rates) converges with research on development and the phenotype, including the behavioral phenotype.

And as we shall see next, the most important developmental and evolutionary strategy of life as a whole, expressed on all levels of evolution, from its most primitive all the way to the human, can be called, adapting Ashley Montagu's suggestive and fitting phrase: "growing young." It is the gradual evolution of involution, actualization and individuation of the power of infinite potentiality, or, finally: evolution becomes conscious of itself in the human as the essence of becoming itself, and this essence—our essence as representatives of life as a whole—is that we learn how to individualize the cosmic, the potential and virtual; to embody what is everywhere disembodied, namely, universal creative becoming, be it through matter, life, mind, or hearts.

Growing Young

It was once more Gould in his highly influential "Phylogeny and Ontogeny" (1977) that brought from oblivion the fundamental problematic of "neoteny" in primates and humans which helped our understanding of evolution as a whole. Gould started by resurrecting Louis Bolk's neglected investigations, and this opened new research and debate regarding neoteny that shows much promise as part of the second scientific revolution and the forming of a new synthesis between social science and human life. Bolk (1866-1930) called his discovery "retardation theory," and summed it up in his classical statement: "Man, in his bodily development, is an embryonic primate that has reached sexual maturity... We represent the primitive conditions of the evolution of our species." (Bolk, 1926) Placed in the whole context of our study thus far, we can say: *human becoming*, understood as evolution becoming conscious of itself, brings to expression nature's general strategy of evolution as a whole. This "strategy" is indeed remarkable, because it makes evolution possible by means of ever recurring phases of involution. Meaning that the more evolved beings are those that have managed to integrate and embody in their bodily plans, organs, and functions, more of the universalizing, neotenizing, characteristics.

We evolve "higher" and further by "going back" to our primitive, less specialized, more open, beginnings, and take our next jump ahead not from the pinnacles of our most complex and sophisticated tips, but from our most *embryonic sources*. This is because those sources are far richer in potentials, and open for adaptive variations and selections, than the already formed specializations of an advanced evolution. This is the reason why we discover that neoteny and related phenomena (juvenilization, paedomorphosis, fetalization) are not chiefly primate or human phenomena, but universal; they show themselves so clearly in human becoming only because they are everywhere present in nature, and as many new researches are finding out, they are indeed far more central to evolution than was previously thought.

Arthur Koestler, a sharp observer of life and science, summed it all up in these fine words:

> Biological evolution is to a large extent a history of escapes from the blind alleys of over-specialization, the evolution of ideas a series of escapes from the tyranny of mental habits and stagnant routines. In biological evolution the escape is brought about by a retreat from the adult to a juvenile stage as the starting-point for the new line; in mental evolution by a temporary regression

> to more primitive and uninhibited modes of ideation, followed by the creative forward leap (the equivalent of a sudden burst of 'adaptive radiation'). Thus these two types of progress—the emergence of evolutionary novelties and the creation of cultural novelties—reflect the same undoing-redoing pattern and appear as analogous processes on different levels.
>
> (Arthur Koestler, *Janus*: *A Summing Up*).

As I was working to add some last remarks to this chapter, recent research came to my notice that illuminates one central detail of neoteny, and provides further evidence of its far reaching significance. The article, worth quoting at length, is "*Transcriptional neoteny in the human brain.*"

> In development, timing is of the utmost importance, and the timing of developmental processes often changes as organisms evolve. In human evolution, developmental retardation, or neoteny, has been proposed as a possible mechanism that contributed to the rise of many human-specific features, including an increase in brain size and the emergence of human-specific cognitive traits. We analyzed mRNA expression in the prefrontal cortex of humans, chimpanzees, and rhesus macaques to determine whether human-specific neotenic changes are present at the gene expression level. We show that the brain transcriptome is dramatically remodeled during postnatal development and that developmental changes in the human brain are indeed delayed relative to other primates...
>
> By comparing the gene expression profiles in human, chimpanzee, and rhesus macaque prefrontal cortices throughout postnatal development, we have found ... a significant excess of genes showing neotenic expression in humans. This result is in line with the neoteny hypothesis of human evolution and provides insight into the possible functional role of neoteny in human brain development. Specifically, we show that at least in one of

the 2 cortical regions studied, the neotenic shift is most pronounced at the time when humans approach sexual maturity, a process known to be delayed in humans relative to chimpanzees or other primates. Furthermore, the neotenic shift particularly affects a group of genes preferentially expressed in gray matter. Intriguingly, the timing of the shift also corresponds to a period of substantial cortical reorganization characterized by a decrease in gray-matter volume, which is thought to be related to synaptic elimination. The developmental pace of changes in gray-matter volume has been associated with the development of cognitive skills among humans (e.g., linguistic skills) as well as with the development of disorders (e.g., attention-deficit/hyperactivity disorder).

(*Somel et al, 2009, www.pnas.org*)

So, perhaps we are really neotenic apes, as Bolk suggested? However, in the new emerging paradigm, this characteristic is fully human and positive, because it can be integrated into the whole new picture of a creatively evolving human in a creatively evolving universe. To be retarded, means, therefore: to be able to evolve much further, because going back, or retaining, the embryonic proportions, functions, and timings, allows the creative faculties to draw on a much vaster potential of creativity. If we are indeed growing young, as Ashley Montagu put it so aptly, then the future of our evolutionary old age may become the future of our most significant creativity. But this depends very much on how we understand and interpret this fact. Montagu was writing in what was in the 1960s a humanistic context, committed to the development of human potentials and the structuring of society and education to fulfill this aim. This is what he said in *Growing Young*:

Based on the evolutionary facts, we may define society as the nurturing life-system that generates and

> extends the neotenous traits of humanity with every generation. The perspective of evolution shows us that our neotenous, extended childhood, our lifelong youthfulness, becomes the single most commanding fact upon which to design all social and productive relations. The child, as Simone de Beauvoir has so well said, surpasses the adult by the wealth of his possibilities, the vast range of his acquisitions, and his emotional freshness. Throughout human history neotenic processes were sustained and succeeded within the evolutionary matrix because social organization rapidly evolved to support the demand of prolonged childhood, to afford the protection, nurturing, learning, and interpersonal support and collaboration essential to the continuing development of human potentialities.

In other words: the human is the most universal expression of evolution, because it has retained a potential element of evolution's virtual, that is, original beginnings, (which are pre-individualized and pre-differentiated) in a biologically individualized and differentiated form. In the human, biological evolution has solved the problem it set for itself, and has managed to bring together, synergistically, the secret of virtualization and actualization, that is, of an individuation that doesn't limit the universal, but precisely makes the universal (or virtual) individualized. Now humans are becoming the creative agent itself, and evolution is becoming self-conscious of its innermost secret. Thus, the human is today on the verge of understanding itself as such, as evolution's solution to its material, embodied, and biological problem. Realizing this fact, the human will begin to address its own future evolution, now that it is going to be free from its organic embodiment. The human will have to decide in which direction it is going to take its future transformations.

Now the old humanism, as truthful and well intentioned as it is, must not stand still. It must greatly intensify its understanding of the human if it is going to be able to "compete" with the above described developments in robotics, genetics, and the latter-day utopias of a scientific-technological "singularity," in which the human will be wholly merged with AI and in which nano-technological genetics and medicine will have extended human biological life to immortality. Let humanists have no illusions about this fact: the human is about to be wholly virtualized, if not in an ecstatic apocalyptic event, then in due course. Therefore, the question must be: Can the human be virtualized *in an essentially humanistic manner*? The event of this virtualization has been upon us since the beginning of the age of natural science and modern technology, and as said above, it is the event of human's becoming conscious of evolution, of evolution becoming self-conscious, and of humans beginning to control, master and use its deeper forces for their own purposes. But what are their purposes really? Is physical immortality the apex of our endeavors and the final solution to all our problems and challenges? This is precisely what we are going to discuss in the next chapter. In whatever way we answer this question, we must first understand that this major modern evolutionary trend is here to stay, and it is indeed accelerating itself exponentially.

There can be no question about it. Humanity is already in the very midst of its irreversible transformation. As neoteny or becoming is established as a major force of all natural evolution, including the human, it will naturally be mastered, used, and manipulated according to our *concept of the human*. Therefore it is high time for philosophers, educators, artists and socially engaged people to take humanism itself into its next level. And this level can only be effective, if we can demonstrate that the human stream of becoming can think through and practically actualize and individualize its involution, de-actualization and de-individuation. This means nothing less than to transform humanism and to virtualize it beyond its present organically embodied state, to transform human cognition, creativity and ethics beyond the body in order to understand how to embody it again in new bodies and new actualizations. If the human is to grow into its own "singularity," it must be able to develop the cognitive tools and faculties, the creative intensity and moral strength that are necessary, in order to become the human of the future. It will depend on our understanding of human becoming as a whole, on our comprehension of the **event** of our time in science, philosophy, history and art.

Chapter Two: The Event in History

Part One: Individuation and Reversal

IMMENSE AND DETAILED KNOWLEDGE has been amassed through empirical historical research in the last century and a half. But this multiplying of information and detail, great and useful as it is, nevertheless, cannot hide the fact that our *understanding* of history lags far behind. As I tried to show in the chapter on the event in science, in this field we find current discoveries that energize our creative inventing and the generating of new concepts. In the field of history the situation is different. It is far poorer in supplying energizing nutrients. This requires from the researcher of the Event a greater reliance on his own forces. The reason for this specific situation is that our research, thinking and living in history, in the political and social fields, is still much more unconscious than in the others. This is caused by the fact that in history we are wholly submerged; we can hardly find a way to "get out" and view it from outside as external observers, as we do in natural science. Furthermore, here we are dealing with the *human being and becoming* in its entirety; while with science and philosophy we may at least believe that we are dealing only with this or that aspect of the human being. History reflects humanity's process of becoming as such. This is the main difficulty in the historical and social fields, which confront us when we strive to develop even the most rudimentary real concepts. In order to make our historical conception "worthy of the Event," we must, therefore, struggle to create new concepts to understand the creature that makes history, namely, the human being. Yet our understanding of ourselves lags far behind our understanding of so many specialized and limited fields; quite apart from the fact that the picture of ourselves, reflected back from history, the "who" we see when we gaze in its mirror, is still so very different from everything we know, or would like to know, about ourselves.

In this first introductory chapter on the event in history I will try to create conceptual tools of operation as I did in the previous chapter. These tools must prove strong enough to deal with the *real itself*, or with "the present from the reality of the present." This can only take place inside the immanent pulse of real human life in history and social life today. But we are just beginning to realize consciously that we are beings of becoming, that we *are* this becoming. This is the meaning of modern history, after all, to become self-conscious as a mature humanity in the Milky Way. But this project is very recent, very young. Therefore, we are still largely living through it half consciously or *unconsciously*, in a deep historical sleep.

In Search of the Ur-Phenomenon of Modern History

My search and struggle for many years in the historical and social fields was dedicated to conceptualizing "the primal phenomena" of modern humanity and modern historical life. I needed it in order to find a foothold that would make it possible to navigate rationally in the dark forests of modern history, in which the paths are often so entangled and misleading. After many trials, I reached an *evental* historical location, the darkest of modern historical sites (which for me was also actualized as real Event when 'I was lost in the middle of life's dark wood,' as Dante said). Here, after a time, I could build a forlorn and lonely, yet actually dependable and faithful, creative home or starting point. It was there, in the moment when I was truly lost in history's darkest time, in which I found myself also in the middle of my individual lifetime's journey, that I was able for the first time to transform and personalize the evental site of history and consciously merge it with my "life's middle." This operation became my individualized operative tool of evental research. I learned how to produce it and how to grasp and handle it with my own forces of life, with my own biography or "personal history." The energy for its production was created by merging together two wounds: the wounded 20th century and my own wound. Then I had to train myself to control and channel the wounds' energy in order to forge from its pain, wonder, and mystery, a sword of effective differentiating and reversing difference. I had to learn to construct, individualize, and actualize it as my own life's primal or Ur-Phenomenon.

It was through using this special operative tool, that I could create an appropriate embryonic, seed-like concept, capturing the sense of the Ur-Phenomenon of modern history. This seed of operating force is therefore *not* an intellectual child of a thinking brain, but a *real seed* composed from the life forces of history's destructive wound and healing fountain. It is a healing power that becomes cognitive if you can capture it in the greatest proximity to the wound, because it is flowing out of modern history's wound and sickness. I gradually transformed and conceptualized it in order, eventually, to let it ripen until it could take the shape of the present concept of the primal phenomena of modern history. And this concept, in its present form, captures the relation between an idea or an ideal and its realization as the primal phenomena of modern history. This concept alone, so I believe, can bring together, both rationally and scientifically, the research of the history of modern ideas with the empirical research of real history, because it brings to light a unique modern difference that prevails between idea/ideal and realization. As we shall presently see, an observation of this **difference** between ideal and real shows, first, that there is a paradox, contrast or **contradiction** between them. Then, on closer inspection, we realize that they are really opposites of each other. The investigation of this contrast and opposition reveals, finally, that a process of reversal is taking place between them. This perception then gives birth to the concept of **reversal** as the Ur-Phenomenon of modern history as such.

Modern Individuation

It is largely accepted that only since the Renaissance have people begun to feel, recognize and identify themselves as having universal human values shared by all, and that in that era, mankind began to become conscious of itself as one humanity for the first time. Jacob Burckhardt formulated this thesis in

1860. This means that the universal as such, variously referred to as God, Nature, or Idea, was grasped by the human individuality *in a personal manner* for the first time in history. The human individual also started to experience herself increasingly becoming emancipated and estranged from external religious and social authorities, and felt that what was formerly transcendental had now become immanent, inward, or even the essence of her human being. As proof of this thesis, historians point out that only as late as the end of the 18th century, did humans formulate for the first time (in America and France) a fully conscious declaration of universal human rights and values that were now placed above race, nation, gender, religion and social-economic position. Together with such declarations, new political ideals were formulated at the same time that fraternity, equality, and liberty were conceived as the condition for a healthy social life. At the same time, natural science was established in its modern, universal form, using mathematics as an international human language to discover, formulate, and utilize "universal natural laws."

The strange thing is: *the more humans become individualized, the more they desire to universalize.* This paradox made itself visible with the discovery and practical realization of the universal both in social and political life and in the universal laws of nature. It is individual people that unearth, utilize, and realize the universal. This means, furthermore that the most universalized ideas and ideals become integrated into an emerging human freedom. The human begins to control these forces and make them her own; and the more each individual is emancipated from her external domain and hegemony, the more she strives to realize them in social life and natural science.

The most essential observation here is that what is individualized is *precisely the universal*. The human individuality feels and experiences herself as a universal human being, precisely by dint of experiencing herself as a private personality. We come to realize that it lies in the nature of the case of modernity itself that the universal is individualized in so many people on one hand, and that the individual is perceived as individual *only because she is rooted in universals*, on the other. But privatization of the universal and generalization of the individual means a *mutual reversal* for both. Universal force and power is put in the service of a personal, private ego, and the same personal ego feels itself enlarged by this into universal dimensions. Transforming universal into individual means nothing less than individuation (or privatization) of the universal, which transforms and reverses the universal and makes it into a personal private possession. At the same time, the universalization of the individual allows the personal self, the ego, to expand its private egotism and make it *universal egotism*. Egotistic universality, on one hand, and universal egotism, on the other, merge together in this double reversal. In other words, modernity means exactly this: placing the content and force of the universal (God, Nature, or Idea) into the hands of the individual, private human personality, gives it increasingly powerful means to control, change, and realize this force in its economic, political and cultural life and to use it in its research, utilization and manipulation of nature. A private ego that controls the forces of God, Nature, or Idea, makes them its own, and uses them for its personal priorities in social life and in nature.

This means that since the private individual is emancipated from external divine and natural universal forces of form, order and meaning, she begins to internalize and control them herself. She becomes ever freer to use them as she wishes. And this is the reason that when we observe modern history and how universal ideas, human values and natural laws and forces are realized in social life, we discover that they have been realized as their *exact opposites.* The French revolution is an archetypal case of all such reversals. It even seems that modern humanity was born to its new self-determination and has ever since consistently created, in all spheres of life, an *exact reversal* of everything it considers just, right and good, and natural. A careful investigation, therefore, of modern and contemporary history and civilization reveals a uniquely modern relation between human ideas and human historical reality, a relation of an exact reversal, which I believe to have discovered and justifiably termed the primal phenomena of modern history.

The dark wood of history becomes transparent on all sides when we observe this remarkable novelty at its center. The human becomes the site of this reversed becoming. The most powerful and universal force of reversing becomes individualized as the key operation in all fields of modern social, political and natural life. It immanently governs the new human constitution from the inside, nay, the human is becoming this constitution itself. It directs, as a constant reversing difference, all the processes of realization and implementation of our human values, ideas and ideals, as well as our social and political principles. Though this fact has obviously been noticed in historical research, it is constantly misinterpreted, because the real reason, cause, or "primal phenomena" governing it is not properly understood.

Failure, Regression, or Reversal?

When most historians speak about the failures of modernism they often use categories such as failure, deviation, transgression or the like as variations on the more fundamental concept of regression to older, pre-modern social and political ideas or regimes. What is usually meant, to put it rather crudely, is that modern humanity is still too weak to resist older and well-entrenched social ideas and habits. In this vein, when liberal researchers acknowledge the 'failures' of modernism, they often place the blame on a conservative 'human nature' that preserves obsolete social forms, and uses them to fight against the new. Conservative historians will often find the cause of failure in the modern capacity to crush the old values and social structures through unbridled belief in progress and modernization.

First let us look closer at the **content** of this notion of "historical regression."

The essential question that must be considered here is—does historical *regression* really mean an objective return to ancient ideal, social and moral contents? On inspection we find that in reality it is not at all a return to older times, but a wholly modern adaptation and utilization of older contents for modern aims, utilizing purely modern means. It actually *revolutionizes* the so-called "ancient contents" in a radical manner. It can be demonstrated in detail how the ancient content is totally transformed through a rigorous process of selection, change and adaptation, that shapes it in entirely new and decidedly modern ways, creating an original, substantial, and modern social form. Even when the conservative, the repentant, the reactionary, believes ardently that he "returns to the pure sources" of his fathers' religion, nation, or social stance, he is using and manipulating his tradition and heritage, as a matter of historical fact, according to his present, modern, ideas and needs. The process is as follows:

1. He chooses his material selectively.
2. He interprets at will what he selects.
3. He invents wholly new elements.
4. He intentionally mixes them with the elements he has selected and interpreted from his tradition.
5. The social, cultural, or political regime that he founds will reflect this really new and original blend of elements in a typical modern (individual-universal) form.

Now this observation of the formative process of these contents leads naturally to a consideration of the **form** of the created synthesis. The form is different from the content only in this regard, that among the elements of the content we do find (though, in markedly transformed manner) elements of ancient contents. But the process of formation and the form of the "return" is actualized in a fully modern way. Therefore, we may say that not only the process of selection, interpretation, invention and realization of ideas (which are taken to represent the purest reflection of original sources) is modern, but also what is more, the *forms of their actualization* are invariably modern. The realization of so-called ancient, purer, contents, will be actualized by using any available modern tool, whether conceptual, scientific, technological, social or political, in order to create a society that returns to "God given values."

If the conservative revolution is properly modern, of course the radical revolutionary activist that fights against the old regime and its regressive and dominating heritage is thoroughly modern. He is actualizing the same process, only with the (relative) reversal of content and form. While the Bolshevik, for example, believes that his ideal content is really new (after all he refers it back to the French revolution, to Marx and Lenin and not to Mount Sinai), the **form** of the regime he creates (the dictatorship of the proletariat), is nothing but a real, though infinitely strengthened, *similitude* of the Czarist regime against which he fights. There is, of course, this modern difference, that his murderous zeal, utilizing scientific-technological means of mass destruction, makes not only the cruelty of Ivan the Great but also the worst tyrants among the Roman emperors and Asiatic despots look like the play of harmless kids. If this is true, then, are not all contemporary revolutions inherently modern? For example, was Ayatollah Khomeini's revolution in Iran (1979) not just as modern as the so called 'truly modern' dictatorship of the Shah, who was covertly supported and controlled by the US, and who actually usurped (1953) the much more democratic regime of Mohammad Mossadegh? Are not all fascist and religious revolutions, which are called 'regressive,' actually just as modern as the communist revolutions? Although it may seem paradoxical, what about the 'progressive' liberal ideal of a global 'free market' dominated by the US? Wouldn't this economic tyranny be one of the most "regressive," because the most hidden, of all modern regimes?

I believe arguing that Stalinism (and Fascism as well as Capitalism) is a regression to older tyrannical regimes only proves that one cannot differentiate between *external similarities* and internal, *essential differences*. This implies that there has been no real comprehension of the innovative modernity of contemporary totalitarianism. The repressive regimes not only use modern phrases about fraternity, liberty, and equality to serve their ends, but above all they function through forms typical of modern consciousness and conduct. They therefore utilize fully modern means in order to realize their goals, synthesizing and marshaling with tremendous force and acumen all available modern technology, politics, and media for this purpose.

What is too readily overlooked is the fact that the radical, conservative, and capitalist revolutions are all equally modern. This is true because all three embody and use the essence of modernism, which is the universal-individual couplet or double bind. The three strive for an absolutely universal totalitarian regime, in which individualism-universalism is preached (but always put into action by making the law of 'state of emergency' or 'state of exception' permanent, as formulated by Carl Schmitt), in which precisely the most modern ideals would be realized, but in their *reversed* form.

Now as I already suggested above, far from being a lack of modernism through a *failure* or *regression*, what is realized through these revolutions is actually *the full reversal* of modernism. But this full reversal is definitely not a *failure* of modernism, nor *transgression, repression* or *regression*, but rather, the purest form through which modernism appears on the stage of modern history. In addition, this purest reversal is realized through both form and content because it operates *beyond* both, and is capable of operating on the purely *virtual* level, from which both are actualized. And this is essentially a modern breakthrough of the human mind to universality, not in any abstract ideal sense, but in the sense which modern science teaches us. For example, the detonation of the atomic bomb is an actualization of the fact that our physical science has penetrated to the universal—and hence infinite—power at the root of all "material" existence. The coming genetic revolution of nature and humanity is possible only because life science increasingly understands the virtual source of genetic differentiation and individuation. The site of reversal is the event of modernity, the universal taken as concrete and infinite virtuality, now accessible for the first time in history to human thinking and human action. This is the reason why all three historical reversals and their variations, insist on realizing their respective ideas in a universal manner, either as "return" or as "progress." It is always through an appeal to universal and global human nature, or rights, needs, and values, as well as universal natural laws, that modernism is realizing its (reversed) goals. In other words, instead of saying *regressions*, *transgressions*, or *failures*, I suggest the concept of reversal with its typical and atypical variations. Therefore I wish to name the three archetypal reversals of modernism's main ideals, those of fraternity, liberty, and equality, as the first universal-virtual reversals.

The source of these ideals is the half conscious *feeling* of the singular essence of nature and human nature. Then they are extracted out of this singular virtuality, turned upside down, individualized and actualized through three revolutions: the French and Bolshevik revolutions, the Fascist and National-Socialist revolutions, and in the continuous Capitalist revolution. The essence of this reversal is wholly universal, and can only be found in the *virtual wholeness of the idea*, expressing the full reality of the virtual. The "threefold" virtual modern reversal takes place through all three revolutions, though with differentiations, variations, and dominances specific to each case. Orwell pinpointed this principle of reversal in 1984 with the concept of newspeak: war = peace, hate = love, slavery = freedom.

Freedom, Privatization, Reversal

Let us start again from the fact that is contested by few today, that the human becomes self-conscious only gradually in the course of history. First, she becomes a private personality, an autonomous agent, only in the new era that dawns in the West in the 15th, 16th, and 17th centuries. Second, this process of "individuation" or "privatization" of the human goes hand with hand with her intense feeling and need that her personal identity should be recognized and respected in a universal way. What is important to notice here is precisely this, that *each individual seeks a universal affirmation* of his rights, beliefs, ideas and social conditions. She does not only demand her private rights, but demands their fulfillment on the ground of their universality, on the account of their being "inalienable human rights" as the American *Declaration of Independence* would call them. The individual, therefore, strives to realize liberty, equality and fraternity as all-embracing, universal human values. But she can only realize those ideals by means of reversing and expressing them as powerful egoistic impulses.

The new era begins, therefore, when the first few humans grasp for the first time (at least with such personal intensity), two contradictory elements operating in one and the same historical moment: namely, their individual personality with its elemental sense of freedom, on one hand, and the universal nature of the human and humanity as a whole, on the other. She experiences, grasps and recognizes herself as a unique individual, emancipated from external-universal forces, which formerly governed and enslaved her through collective religious, philosophical and social beliefs and structures, yet again, paradoxically, at the same time she grasps inwardly, individually, the very same universal forces. We must grasp and delimit this twofold moment very sharply: she experiences herself for the first time as a creator of her thoughts and ideas and she acts individually upon them. This gives her an enhanced sense of individual identity, value and freedom. At the same time those very thoughts and ideas that come together with the scientific discoveries of universal laws and the discovery that the earth is one globe shared by all humans, speak to her about the universal character of human nature, with its rights and values. This discloses a human essence beyond race, nation and belief, equally shared by people who are gradually individualizing all over the globe. Then she also begins to strive to create, through her elementary impulses, a social order that will reflect this new individual-universal experience. She conceives social ideas and ideals and believes, with best intentions, that they would harmonize the growing sense of individual identity and freedom with the equally intensive experience of the universality of human nature and natural laws. But when she actualizes them in social and political praxis, they are immediately reversed.

The Basic Cognitive Structure of Reversal

Let us observe now closely how the reversal is actually taking place. An important cognitive characteristic of the Ur-Phenomenon of modern history is the following. There is no "time interval," physically, psychologically and historically speaking, between the ideal-virtual feeling of the universal modern singularity, in which the universal idea is rooted, and its individuation as an actually realized reversal. As a matter of fact, a more radical statement would be: modernism = individuation = reversal. If we wish to be scientifically exact, we *cannot* say: the modern universal-virtual appears **first** in full consciousness and is reversed **only later,** through the social and political praxis. If this were so, we could point to a place and time, to an individual thinker and social process, which could be identified somewhere in the 15th to the 19th centuries as "the thing in-itself" in its pure form. That is, contrary to common opinion and also to many historical conceptions, it is *not* correct to say that revolutionary ideas are grasped as inspirations that are clearly thought through, expressed consciously, and only later falsely realized (through failure or regression) in the revolutionary social process. But the form of this modern consciousness is the form of individuation. Hence the *form itself* only allows them to emerge reversed *at the original roots of their appearance.* This transformation occurs precisely *before* the moment they rise to consciousness. Universality has *already* assumed its reversed, individualized form *in the very moment* ideas are individually grasped.

For example, if we study the form and content of the ideas of the thinkers of the French revolution, including Rousseau and Voltaire, we will find that the ideas they conceive and formulate are *already thoroughly reversed*. The cause of the reversal in Rousseau is found in the fact that he excludes the free individual from the general will, while in Voltaire it is found in his fanatical hatred towards God, church and royalty. Taken together, they supplied the revolution with impulses that led to the dictatorship of the proletariat (Rousseau) and the extermination of the ancient regimes (Voltaire). They gave the revolution a reversed ideal foundation, which was then implemented politically with radical ferocity.

The moment of reversal is really unconscious, that is below or above the wide-awake level of modern self-consciousness. After all, self-consciousness is precisely **my own** self-consciousness, because of the fact that it is individualized in and through **my own body**, with its brain- and sense-bound thinking. It is the body that makes thinking, which is a universal, non-organic, infinite virtuality, into my own, and hence reversed, appropriated, individual possession. Until the latter part of the 20th century, we cannot locate such an event in modern history because the state of consciousness that would allow humans to grasp consciously virtual, singular, pre-personal realities is only just emerging. True, many experience "intimations of immortality," but they experience them in a nebulous, dreamy and half conscious manner. But when they begin to implement them in social life, they realize, once more, that the best intentions have created evil reversals in real life.

That is: the universal-virtual is experienced vaguely as sentiment, feeling, or mystical intuition, but the modern mind strives instinctively towards full consciousness, that is, for a clear cognitive grasp of concepts and percepts and the forming of scientific judgments. The reversal happens instantaneously, and the virtual that is felt is placed into fully clear but fully reversed concepts and judgments. The actual physical, social-political, cultural and economic reversal of the ideal happens first in the transition from the twilight experience of universally felt intuition and sentiment, to the form of clear, individualized, modern thinking.

Would it be possible for real individualism to be actualized at all if such a fundamental reversal would not have taken place? Could or should it have been otherwise? We can perhaps conceive of a reversal of individualism, of its singularization and virtualization, but we can do so only because the individual is already here, it has first become an historical fact! The radical novelty of this transformation is so immense, that it seems rather impossible to assume that it could have been achieved otherwise. Reversal must come first, and *its reversal* second. Yet, again and again, we are surprised to find that it is so, and tend to see the reversal as the exception rather than the rule itself.

In the moment of the birth of modern self-consciousness and individual identity, in the moment in which the individual determines her value through the universal, and the universal is appropriated and used to ground the individual, is it really thinkable that a reversal of both could be somehow avoided? If it were avoided, then obviously no modern era would have started at all. The divine, or nature, would have continued forever to reign supreme, as in the Middle Ages and the Ancient cultures. After all, this moment is the very first in which the *universal as such* is grasped and realized by the self emancipating individual. At the very same historical moment, the individual is freed from the external controls of this universal, and begins to command it by his own, now increasingly and irreversibly individualized forces. Therefore, objectively speaking, we must assume that no external power, be it divine or natural, will now intervene to prevent the individual from (mis)using the universal infinitely, in the most ego-tistical manner, in order to augment his private power and to fulfill her personal desires. Indeed, individuation would be unthinkable if the beneficial universal forces had not stopped their external guidance, in order to allow human freedom to evolve in modern history.

Now we are in a position to lift to consciousness what, in the first 300 years of modernity, has remained largely unconscious, namely, the cognitive structure of reversal. If we want to move forward in social life in the 21st century, it is truly necessary to lift this structure into the light of full self-consciousness. Only through this light can the unconscious reversal be consciously counter-reversed. So how does the modern reversal *take place and take time*?

Well, it is taking place right at the source and from the very beginning, in the very moment in which the private, self-conscious soul and mind, grasps a universal impulse, idea, or ideal. Is it not the private person, "my private human subject" that supposedly grasps the purely universal? Does the primal act of self-conscious cognition, termed in Phenomenology, *eidetic intuition* or *seeing of ideas*, not immediately cause a reversal, because my "I" (as private subject) grasps the "idea" (which is universal)?

Let us pause a moment to reflect. How do we grasp an idea in reality if we are not encountering something *and its reversal* in one and the same act? Each thinker striving to overcome Kant found a uniquely constructed philosophical trick to merge and cancel these contradictions according to his own specific bias. After all, as Kant, Schopenhauer, and Eduard von Hartmann, and many other old and neo-Kantians have said and will be still saying in all times to come: is the universal not **by definition** universal, and hence not individual, and therefore can never become individual? Is the individual not **by definition** forever individual, and will never become universal? Therefore, logically speaking, will not an abyss of real contradiction always separate the two? Who will dare say that he can overcome this seemingly eternally justified objection?

But logical definitions aside, my "real grasping I" is my personal "I," grounded in a fully private animal-human subjectivity, which makes me what I am as a human personality. Even if I could grasp an idea, I would have to do it with an "ideal part" of myself. But this would only mean that we transfer the problem to a different place. I would have to explain that in me, in my subjective self, there could be such a non-personal, non-subjective, ideal site, with a universal edge that could truthfully grasp objective, universal, ideal content. I must therefore imagine the human to be a twofold, cognitive amphibian, on one side, subjective and wholly human, on the other side, universal and semi-divine; but then where and how exactly do the two sides of my double humanity meet in my human being? It is clear that to locate a meeting place that would bring together the personal and universal inside my human personality, would only mask the problem and not solve it, because now I would have to explain how exactly the two opposite halves of my own personality could meet each other in my own being. Unfortunately, this explanation will soon prove to be a logical and more than that, an existentially real impossibility.

But let us assume for argument's sake, that I am somehow "endowed" with this divine faculty (as Plato and Aristotle could still believe), and that I have this edge, this *Nous* or *active intellect* as an integrated "part" of myself. But even if this could be admitted, then, in the next moment, after the eidetic part of me grasped an ideal content, the other part in me must immediately and instinctively transform it through the psychophysical connection with my own private-subjective representations, as well as my feelings, desire and lust. This means that I have absolutely "no time" to hold the ideal content in pure inner reflection, because (again, let us assume this only theoretically) when "I" grasp the ideal, "I" have already immediately made it into my own private possession. I have already realized it as my own representation, formed according to the most subjective feelings, desires, habits, and so on, of my private personality. In short, I have immediately reversed its universal essence into something entirely subjective and human.

I don't think that I have to prove here that when I observe my active mental, soul and bodily life, and study my own expressions, in thinking, speaking and doing, that it is my subjective nature that acts, that my personality is the only source of my actions. If it is not so difficult to prove that I either do not have eidetic perception at all, or, in the case that I miraculously have it, it is immediately privatized and reversed, then it will be really quite unnecessary to prove that the personal expressions of my human personality are nothing else but my personal, subjective, expressions. What else on earth could they be? Let us say I have truly grasped an idea and subsequently made it into my private possession. Now I wish to communicate it to you. Obviously I will only be able to communicate to you what I have made my own personal possession, that is to say, only the reversed ideal content. What would have remained intact as a pure universal, objective, residue, from this ideal content, after it has been transformed through and through inside my subjective being? After being taken down and filtered out by all my subjective levels of being, from my private thinking and feeling, down to my subjective desires, habits, norms, education, cultural and social adaptations and accommodations—what could be left?

Let us say that I truly manage to grasp an ideal part of an ideal, external, objective world. Then, as a matter of fact, my psycho-physical, mental, social and cultural *metabolism* immediately takes hold of it, grasps it, plucks it from the universal existence and separates it from the objective world. "I" then break it into pieces in my soul mouth, eat, swallow, and then totally metamorphose and assimilate it into my own soul substance, down in the depths of my unconscious soul metabolism. What remains and returns to the external ideal world but the excretion? Would this not be the substance and form of the idea I have communicated or the moral ideal I have realized? Therefore, I believe that it must be stated very clearly that the origin of all communication and expression is to be found in the above described unconscious process of metabolic soul privatization and reversal.

The term "subjective expression" is simply redundant because "expression" means nothing but the product of a human subjectivity and not of a divine being or even a being endowed with a "separate intellect," as Aristotle or Thomas von Aquinas could still feel justified in assuming that humans can possess.

Therefore, any expression means always an expression of the fundamental structure of reversal and of a specific reversed content, of a world transformed into human subjectivity, expressed upside down and outside-in. This is again the reason why I believe this reversal is the Ur-Phenomenon of modern history, of modern humanity as such, because modern humanity is the grounding epoch of subjectivity, personality and self-consciousness. Now, this could be experienced, from an idealistic point of view, as tragedy and even a cause for mourning a "lost paradise" or "golden age" of reason or truth. It is therefore necessary for me to state that this is definitely not my point of view! While the dangers of this reversal will be emphasized later on, this statement must be made right now, without which the whole narrative of *the event in history* will be fundamentally misunderstood: the reversal is not only an inevitable evolutionary event, but it is **the most positively productive event in the evolution of human consciousness**, as the major step in individuation. For without this reversal no freedom would be at all thinkable, let alone realizable. And this is, before and above everything else, the real essence of the Ur-Phenomenon of modern history.

Indeed, truthfulness and clarity will be actualized in human history only if this Ur-Phenomenon will be understood in reality. Of course, one could object and say that the reversal, as it takes place historically and in daily social life, is also the cause of the most radical evil. This objection is fully justified, and I will show below, to what extent it is justified. But to believe that, historically speaking, humanity could have been in a position to both comprehend the reversal at the moment of its first historical realization, and avoid it, or counter-reverse it immediately, would be wholly unrealistic; however, again, this doesn't mean that the horrors of modern history are morally justified. And this is the reason why *today* is the time to understand the reversal, to learn how to cancel it, or better still, to learn how to reverse it as soon as it takes place, and not a day, decade, or century from now. Therefore, consciously reversing the reversal is and shall be *the art and science of true historical becoming.*

Part Two: Historical Reversals

The First Beast

NOW WE WILL LOOK somewhat closer at the three revolutions: the communist-Bolshevik, the fascist-Nazi, and the perpetual revolution of the dominant American global market. Jacob Talmon, in his excellent book, *Myth of the Nation and Vision of Revolution,* covers the communist and fascist elements, but conveniently fails to mention the American hegemony. It is not because the American reversal is essentially different from the other two, but because academic Anglo-American historians typically overlook the complicity of their own nations, just as Alain Badiou overlooks the reversals of Lenin and Mao. This threefold reversal is the first fundamental variation through which the Ur-Phenomenon of modern history has been realized historically in the course of the last centuries, only ripening fully in the 20th century. The reversal is as modern as what it reverses and can take place only from the historical moment, described earlier, in which the individual as such and the universal as such appear for the first time in individual and universal human life.

The modern archetypal reversal, expressed through the American and French revolutions, demonstrates the above thesis of reversal as Ur-Phenomenon of modern history. We can grasp here very clearly how it works, and understand the latter social and political developments of Bolshevism in the east, Fascism and Nazism in the middle, and Capitalism in the west, in the course of the 20th century as the evil metamorphoses of liberty, equality, brotherhood. These three brothers are like three "heads" united together in the body of one and the same "beast" of historical reversal. As the story goes in chapters 12 and 13 of *The Apocalypse of St. John*, after the first beast completes its mission, to show us for what we are, as despicably evil and fatally mortal, there comes the second beast, which has the "healing" task: to demonstrate in a practical way, and convince humanity, that the "wound of mortality" inflicted on the first beast (the wars and evils of the 20th century), can be miraculously healed. And then it sets out to organize all aspects of human life in a wholly utopian and revolutionary manner. Both beasts represent the *historical expression* of the reversal, and its *fully materialistic creativity*. Let us concisely portray the three heads of the 20th century's beast of reversal through representative quotations, each demonstrating the positive contribution of east, west and middle, each of which suffered its typical reversal.

The first one is a representative of the good future forces that the east as a whole and Russia in particular, has to develop. It is used often by Levinas, a Lithuanian Jew, whose philosophy as a whole represents in the most fine and elaborate manner this "eastern" positive mission of the future. It is taken from Dostoyevsky's *The Brothers Karamazov*: *"All of us are guilty for everything and before everybody, and I am more than the others."* It is not difficult to see how Bolshevism, in its Leninist-Stalinist as well as Maoist forms, is the exact reversal of this sentiment. It says in effect: *"All of you are guilty for everything and for everybody and I am more innocent than all the others."* So the practical consequence is naturally, that I have to be the first to kill everybody else, not as self protection, but as ontological murder, because the other *as other* is primordially guilty and his guilt has already being established and needs no proof. This is the cold, annihilating reversal of the east that was actualized through the Bolshevik beast in the 20th century.

Next to the Russian reversal in the east, we have its twin reversal in Germany, that is, in middle Europe. Many have raised this deeply unsettling question: how did the nation of *Dichter und Denker* (thinkers and poets) become a nation of *Richter und Henker* (judges and executioners)? How did a place like Weimar that in the 18th century was renowned for the most human-universal life forces of the Goethe-Schiller age, become home to the concentration camp Buchenwald? Only the theory of reversal as presented here can explain this. For the most beautiful and fruitful seeds of future spiritual life, science and art were created by Goethe and his contemporaries. Just consider his natural scientific method (only beginning to be appreciated today as a real holistic alternative to modern science), which stresses the investigation of *Urphenomena*, or the new image of human development in *Faust* and in the legend of *The Green Snake and the Beautiful Lily*.

For example, Goethe expresses this new image in the following initiatory ideal:

And until you have grasped
This: Die and Become!
You are just a dismal guest
On the dark Earth.

In order to better understand the German reversal, in this sense of *Stirb und Werde (die and become)*, let us bring it together with Germany's original myth, the initiation process of the god Odin, as given in the *Edda*:

I know that I was hung on a tree
Exposed to the winds
During nine nights
Wounded by a spear,
And I was sacrificed,
I, Odin,
By myself,
For myself.

Here the original middle European "I" initiation is portrayed in all its essential elements and meanings, and note that it is the "I" that sacrifices its lower aspect for the development of a higher self. When this is reversed, immediately the meaning of the National-Socialist reversal becomes clear: instead of the spiritual-moral sacrifice of the lower "ego" in order to become the higher "Self," they reverse it and sacrifice the higher self in order to preserve the lower. The national race, blood and soil ego is elevated into universality, while the universal, represented by the Jew, is exterminated. German history in the 20th century *could have* consisted of self-transformation, *Stirb und Werde*, which would have brought healing and love to the world. Yet this reversal of Germany's destiny also became an *inverted* killing of one's true, higher self, and hence a moral self murder, a national suicide, a most exact mirror image of a true initiation of the "I" which it is middle Europe's mission to fulfill.

Side by side with the eastern and middle historical reversal, we have the western Anglo-American reversal of the *true and beneficial mission* of the English speaking peoples. As masters of the world by destination and capacity, first the British Empire, and now the American, could and should have used their given power and fantastic creative physical abilities, to create affluence for humanity as a whole. Instead of this, we see a complete reversal. Georges Bataille in *The Accursed Share* (1949) has defined the western catastrophe in its economic practice (which is served by its science, technology, education and culture as a whole, all geared to this one goal). He says that the problem of the world economy, ruled by the US, is what you do with the "excess" of economic prosperity as such? *Production as a goal in itself* can only become as social disease, a real social cancer, because it will not annihilate itself through free giving, leading to monstrous accumulations, untold riches and power in the hands of very few, and also to necessary cycles that will destroy this excess. Economic life therefore makes *uncontrolled destruction of its surplus (= war)* a necessity, because unlike ancient societies, it has not institutionalized "free giving," as gifts, such as the American Indian Potluck ritual (Bataille was using the studies of Marcel Mauss, *The Gift: The Form and Reason for Exchange in Archaic Societies*). Bataille put is so:

> I will simply state, without waiting further, that the extension of economic growth itself requires the overturning of economic principles—the overturning of the ethics that grounds them. Changing from the perspectives of restrictive economy to those of general economy actually accomplishes a Copernican transformation: a reversal of thinking—and of ethics. If a part of wealth (subject to a rough estimate) is doomed to destruction or at least to unproductive use without any possible profit, it is logical, even inescapable, to surrender commodities without return... the possibility of pursuing growth is itself subordinated to giving: The industrial development of the entire world demands of Americans that they lucidly grasp the necessity, for an economy such as theirs, of having a margin of profitless operations. An immense industrial network cannot be managed in the same way that one changes a tire... It expresses a circuit of cosmic energy on which it depends, which it cannot limit, and whose laws it cannot ignore without consequences. Woe to those who, to the very end, insist on regulating the movement that exceeds them with the narrow mind of the mechanic who changes a tire.

Purely free giving is not only an ethical ideal, but also an objective economic reality. After all, pure consumers, that don't produce physical goods and services, can only exist through an economic free gift: children, our next generation, for example. But any *creative activity* that will become fruitful in the future must be provided today by means of a surplus taken from the natural, given, excess produced in the global economy. Rudolf Steiner described this as a fundamental law of economy in 1922 in his *World Economy* lectures. Now, as history's fable tells us, this third beast, or third head of the first beast, fought a war of life and death over the domination of the earth with its two other sisters, or heads, the German and Russian, and has won the three world wars (the cold war was a real war for all purposes). This victory was celebrated with the collapse of the Soviet Union, and the establishment of the US Empire in the 1990s. Since then the US has been consolidating its global hegemony. But this victory can be seen also from another perspective: it is gathering all the fruit into one basket, which is also gathering the three reversals into their original united root in the body of the first beast. And this reassembling of the three back into one, allows the one to appear in its true, pristine, light.

The 21st Century and the Second Beast

If the reality of the 20th century reversed the three great ideals of the 18th and 19th centuries: *liberty, equality and brotherhood*, the ideal of the 21st century is the reversal of humanity's eternal dream of achieving immortality. Of course, this has been the ideal of western, scientific humanity since More and Bacon's utopias. For example, in his *New Atlantis*, Bacon portrayed the first clear vision of the new scientific ideal to totally transform the human by using purely economic, scientific and technological means. This principle is really the same old eastern religious ideal, now clothed in a western scientific form—*the longing for personal immortality*. This could be called the ultimate reversal of universalization and the infinite intensification of egoism. Therefore its aim is nothing less apocalyptic than the religious ideal, only the means of achieving it are different. It is still the desire for personal survival, only now here on this physical earth and not in a spiritual heaven. Furthermore, the means are as visionary and utopian, namely, the total utopian transformation (indeed transubstantiation) of the human by means of materialistic, scientific and technological forces.

What appeared to be divided into three different and competing ideals in the 20th century seems now to share a deep root of materialistic and utopian belief. After all, the Nazi's belief in race and blood is as radically materialistic as the Bolshevik's materialistic conception of history, society, state and the human being. The economic empire of capitalism is obviously as materialistic as the other two; in this sense we can say that the victory of capitalism in the 20th century allows the global victory of the second beast in the 21st century. For the second beast clearly envisions the utopian ideal and has all the means to realize it in all fields of human life on earth.

Healing the Wound of Mortality

By decoding the complete genome, scientists will understand the letters that make up the words and sentences of the grammar of life. But what is far more important, they will also then understand the virtual source of differentiation and individuation. This means that humanity will be able to understand, control and manipulate life, not at the level of letters and words, or even sentences and paragraphs, but at the level of the *virtual totality of life's potential.* The first thing they will try to do is, of course, "cure" the basic "flaws" of life, starting with birth and death, illness and retardation.

To cure life's original self-limitation, primordial biological wound, namely, *mortality* means the end of the human as we know it, the final chapter in human biological evolution and the beginning of a totally transformed and non-organic evolution. In their vision, the immortalized, post-biological, and technologized human, who will also survive the death of planet earth, will continue to populate the universe and evolve to ever expanding forms of cosmic intelligence. This would be the ultimate scientific victory over religion, a fully materialistic return to paradise, because science would have used the fruit of the tree of knowledge, the fruit of the sinful mind, in order to gain access to the tree of life, and united the two in order to achieve eternal life. This means an overcoming of God's judgment, banishment and protection, and his warning that if this union was consummated in this form, the human would become an offspring of Lucifer, that tempting serpent, for which knowledge, immortality and infinite egoism are one and the same.

The technological virtualization of life's script means that the eating from the tree of life can be put into service of each individual's desire for immortality, and that the primal wound, that made us human for as long as we still have to develop our morality, love and responsibility for each other, will be permanently closed. The wound will be healed, and this "miracle" will tempt humanity in the most potent and irresistible ways, leading mankind to put complete faith, in the power of the beast (keeping to the imagery of *Revelation*) who now can make them immortal. This would transform every aspect of social life. Such a new economic, political and cultural ideal would then create the ultimate, most exact reversal of human evolution: the reversal of the *meaning and goal of evolution itself* in its real essence, the real apocalyptic reversal of the human being on earth. The great utopian ideals, for example, the *singularity* of Kurzweil and *trans-humanism* are expressed most clearly, and are being practically realized with all the messianic zeal and expertise that was previously seen only among the most devoted religious believers, churches and sects. Among the believers and supporters of the growing movements of trans-humanism and singularity, we find some of the best, brightest and most influential scientists, entrepreneurs, politicians and business men and women. Whether or not the singularity will really be achieved in a few decades is less essential, the important matter is that it is already *thinkable* using the scientific and technological concepts and tools that we already possess. It may well be, as many opponents contend, that Kurzweil's prediction of full singularity by 2045 will be proved at least partially wrong, but what is far more important is that the idea of eternal reversal, of *infinite egoism*, is at hand. What is more, the call is strongly resounding throughout the globe, to gather all humanity to wonder at the miracle, and to implement the healing of the wound of mortality in all fields of economic, political and cultural life.

Reversal of reversal: Singularization

In the present historical moment, **the individual is becoming a source of new virtual creation**, either through advanced technology or through self-actualization as a free and creative individual. We are talking here about a **real singularization** (the complete opposite of Kurzweil's physical-mechanical ideal), the creation of cognitive, moral and spiritual immortality. The individual will become an *agent of becoming* and will do so in a fully human, non-technological, way.

It is indeed also the production of a new human, but this new human will be the virtualized essence of the existing human. It is also an *immortalization* but this one is brought about by purely human forces, that do not simply "retain" but infinitely intensify all human faculties, not only intelligence, but also freedom, love, care and creativity. It virtualizes the human, and makes her *cosmic*, but in an essentially human way, rather than through AI. Science has discovered the field of virtual becoming and has investigated the variations, possibilities and infinite technological intensifications of the merging of the human with super human AI. But it will not of itself show us the right cognitive, creative and moral actualizations. This must be sought in modern thinking and modern creativity.

This has been the main subject of post-structural philosophy since the middle of the 20th century. What characterizes the *Event* in philosophy and also in art is the *beginning of the reversal* of this modern historical reversal. This is the first expression of the development of a new form and level of consciousness and being, one that can both perceive and think the virtual as it is, and implement it as such individually in social life. But this new impulse is only slowly beginning to emerge and develop among individual humans since the end of the last century.

Now, indeed, this is precisely the paradox of modernity as such, that only what is grasped in virtualized, singularized, consciousness and expressed through pure virtual thinking, will, and love, can be actualized *without* reversal. This means that it can be individualized in such a manner that the universal will not be egotized, but that the individual will be universalized. This would not be universalized egoism, but a being that is universal and singular—a creative individuality, capable of creating and actualizing real happenings on the level of the virtual with human means. Only such a fully singularized, reversed reversal would be socially fruitful. But:

> "It takes so many thousand years to wake—so will you wake for pity's sake?"
>
> (Christopher Fry, *A Sleep of Prisoners, 1951*)

Chapter Three: The Event in Philosophy

THIS CHAPTER ADDRESSES the same problems raised in the previous ones. The first problem is that natural science opens up a way to appropriate and utilize the virtual in all fields of life. This is the field of non-organic life forces, energies and elements, which includes also the forces that until now were not consciously controlled and manipulated by the human's personal mind and egoistic desires: the forces that bring about birth and death in nature as a whole as well as in the human body. Philosophy must be able to tell us, using the forces of creative human thinking, how science's challenge should be met. It is for this express purpose that contemporary minds must think the Event, because the Event of becoming is the entry portal to the next, higher level, not only of body free intelligence and life forces, but also of impersonal, true love and compassion. There is no other answer to science and technology's challenging achievements, which in their right place deserve our highest recognition and admiration, but learning to think and create outside of the confines and limits of the body- and brain-bound cognition of the past. But outside the body we come into contact not only with infinite intelligence and infinite life, but also with infinite wisdom, love and responsibility for the other and for the earth.

Therefore, it is natural that contemporary philosophy leads us in both directions at once: *out of the body* in thinking the virtual as real, that is, non-organic life and intelligence as cosmic realities, and at the same time *to the encounter with the other* as a real event. Only when philosophy becomes creative activity in this double sense, can we answer the appeal resounding from history and social life for a real reversal of modern history's reversal: are we going to individuate, actualize and singularize the virtual *without* reversing and appropriating, privatizing and egotizing it? The art of individualizing and actualizing virtualities, singularities and multiplicities is not only an entirely new art of creative thinking, perception, cognition and consciousness, it is also a wholly new political, economic and social art. It is a new art of grasping and actualizing the event in and through the encounter with the other. This is the first entry point we have for a real historical transformation, a beginning of real social-historical processes of becoming.

Many thinkers in the 2nd half the last century thought through the event. For example, after decades of joint productive work, Gilles Deleuze and Felix Guattari made the following statement in their last book, "The sole purpose of philosophy is to be worthy of the event." This expresses a profound shift of the whole nature and practice of thinking. We can even say, without exaggeration, that it expresses a real reversal and inversion, inside out and outside in (the German language expresses this in one word: *Umstülpung*), of what used to be understood and experienced as thinking in the modern age as a whole.

But in order to understand this appeal, we must *practice* it. It will little help to try to prove or disprove it, intellectually. Instead, I will endeavor to practically demonstrate in this chapter, that if thinking is not merely directing its forces to conventional philosophical proving and disproving, it retains an *excess* of force that can be made useful in a productive manner. And it is this extra force that allows it to actualize the "event." To actualize the event means to break out of the habitual, brain-bound thinking, exit the representational Cartesian "self," enter the non-organic life fields in the open cosmic worlds, build oneself a virtual individuality, brain and body, as a support for further scientific research and creative production out there, and come back safely with transformed capacities and conceptions. Actualizing the event in full consciousness requires, in addition to the research of the new lands and territories in the open world, a safe landing and re-territorializing back to the body of a new earth and new people. That is, we must demonstrate that the virtualization of thinking, becoming actively engaged in the event, leads back to the body, the earth, the other, and that we come out of the virtual enriched with transformative, life-giving, social, political, educational and healing forces, that embody the fruits of scientific research of the event.

Below we shall offer some examples of intensively productive forces and elements that can be extracted from modern thinking. Because these forces must be still fresh when harvested, they really cannot be older than about half a century. Therefore, our harvest will be an exemplary extract, taken from some of the best thinkers in the 2nd half of the last century (while so many other great and significant thinkers had to be left out). Harvesting their intensively creative productivity is an essential component in the whole effort to humanize singularity (or singularize humanity), as a foundation for a new composition of science and research of the Event. It is also the first step in developing a new scientific method adapted for the 21st century, in which the research, individuation and actualization of the virtual becomes absolutely necessary. This "science of the event" will be presented in my forthcoming book on Cognitive Yoga.

Deleuze and Deleuze-Guattari (D&G)

The necessary productive elements for the present work of composition are extracted from five French thinkers (a younger Italian co-thinker, Giorgio Agamben, must be left out for now). This list is composed of, Gilles Deleuze (including his work with Felix Guattari), Michel Foucault, Jacques Derrida, Emmanuel Levinas and Alain Badiou (may he be blessed with a long and productive life!). They constitute what I call "the holey pentagram," borrowing the term "holey" from Deleuze & Guattari's geophilosophy, where it indicates hidden, underground holes and tunnels dug not only underneath what is territorialized and striated, but also beside nomadically smooth surfaces, as something other than both, drawing lines of flight in the depth of a shattered and devastated earth of the 20th century.

This "pentagram" of thought production has defined the main trajectories of post-phenomenological and poststructuralist thinking. They also continue today to shape and activate fields of new thought, in which thinking and the human are explored and conceptualized. (They are of course surrounded and supported on all sides by a vast cloud of forerunners and contemporaries, such as Bergson, Sartre, Levi-Strauss, Merleau-Ponty, Blanchot, Bachelard, Baudrillard, Lacan, Virilio, Serres, Agamben, and many others). Each one of the chosen five has in turn created multiple conceptual philosophical personas through his interactions with the others, and in each one of these interactions, syntheses, and often violent clashes, I found new materials for my own work of composition.

Let me add that the style of presentation will be shaped in such a manner that it will be possible to experience beneath it a gradual intensifying of productive energy. I would ask the reader therefore, not to focus her attention on the conceptual content alone, but more on the *energy* produced through the intensification of the activity of thinking itself, because this is what the work is all about. It outlines a process of virtual (in the Deleuzian sense of 'ideal without being abstract, real without being actual') composition, which is a creative, even artistic, project.

Now any composition needs a "construction site" as well as "launching site," or the best "place" and "ground" in which it can grow and develop under the most favorable conditions. The site is both a foundation as well as part of the project itself. This site is made of Deleuze's own philosophical work, intensified through the composition he created together with Felix Guattari. Together they achieved a potent and creative hybridization (marked below as D&G). Their work is arguably the most creative and fruitful philosophical impulse in modern thinking in the second half of the 20th century. They are therefore chosen to play the role of "the site of composition" in which the Event in Philosophy is composed and launched.

D&G are gifted in inventing and using a vast array of cognitive tools, machines and conceptual operations. For our present project they will prove particularly useful because of their special expertise in:

1) *unhinging* the thinking of the event from its bodily- and organic foundations
2) *navigating* with its freed forces in the streams of non-organic, immanent cosmic life, and working creatively on this plane
3) *re-actualizing* the event back in the world of ever better physical, social and cognitive forms, that become increasingly more transparent and expressive of the event.

This is the reason why their work holds a key to the challenges posed by science and history: they are the first, in our present time, to have performed the reversal of the reversal, or counter reversal, and have developed the cognitive capacities to work outside of the body- and brain-bound intellect. Therefore they provide a powerful support for gaining the necessary activity, intensity and acceleration we need, the energy of the "lines of flight," in order to facilitate the release of thinking from its embodiment in the organic body and its individuation in psychological experience. This intensity then continues to operate in the working of these released forces in the open, non-organic, virtual life fields of the cosmic world. This is D&G's specialized capacity. They provide the Event and Philosophy with a launching site, that supplies it with useful materials and energies for the complicated and even dangerous take off and safe coming back.

In other words: D&G developed a technique to capture and master the forces that actualize the event in the physical world of time and space (which includes, first and foremost, our human cognition and social-political productions). Then, to reverse their direction, release, liberate and lift them out of the physical world and its forms, and utilize them to think and actualize the event itself in its virtual essence. This takes place on what they called, variously, the plane of immanence, consistency, infinity, and non-organic, cosmic life forces, in order to re-actualize them in new physical, social and cognitive individuations and forms.

As in the last chapter, I will call this key operation as a whole "singularization," which could be compared to Spinoza's reciprocal, mutually interactive *Natura naturans* and *Natura naturata* individualized. It encompasses and operates three distinct activities: "de-actualization" (also called counter-effectuation), "virtual actualization" and "re-actualization" (expression). De-actualization refers to a praxis that liberates the event, with its forces of life and cognition—the life of cognition and the cognition of life—from their embodiment and entombment in the physical world. Re-actualization describes the expressive process of the conscious reincarnation of the event, creating and forming new enveloping forms of actualized expression, embodiment and individuation. Virtual actualization is the purest power of actualization, lifted out of the body, brain, social and cultural forms, and practiced, not in the world of finished forms but on the plane of consistency-immanence-infinity. It is an actualization and formation process of the virtual event of becoming, performed by the freed life/cognitive forces in the open cosmic and infinite world of non-organic life.

De-actualization and Singularization

De-actualization refers to the release of the event of becoming and its life forces, energies and elements, from its embodied forms, habits and situations. In the physical world of space and time the event of becoming is incarnated, entangled and bounded in enclosing fixed forms. De-actualization, or counter-effectuation, refers to the release of actualized and realized forces of becoming, creativity, creation and production, on whatever level of existence, be it matter, life, consciousness, historical events and social-cultural situations, that are formed, hardened and dominated by the powers of form. (This was also called 'The judgment of God' by D&G). This refers to any finished and fixed form, whether physical, biological, psychological, intellectual or historical. But more importantly for our project here (though this is more difficult to understand), is the fact that de-actualization releases not only various realized forces and substances and with them the event, but it releases *the power of actualization* itself from the bonds of realization. Therefore we intend to capture this power, because it is the deeper and more intense power of becoming, and we are going to thoroughly individualize and singularize it through the three-folded practice of de-actualization, virtual actualization and re-actualization. I term this practice as a whole "singularization," or also "actualization of actualization," since it is the operating engine of this threefold process.

Singularization engineers actualization in three modes. First, it reverses the usual, unconscious, actualizations, when it operates as de-actualization. Second, actualization is enacted as a wholly virtual capacity on the plane of immanence, as *virtual actualization*, and then it is reversed again in expressive re-actualization in the physical world. Singularization, therefore, refers to the core capacity and technique that controls all three modes of actualization, because the three require individuation of the full spectrum of actualization's metamorphosis, with its recurrent reversals.

Now de-actualization, when studied closer, releases these elements from their embodiment.

1. It releases the event's actualized envelopes and forms—cognitive, social, or historical with their specific structures and habits, through which the event is incarnating. The envelopes are not the event itself, because the event remains always in a state of pure immanent potentiality and virtuality and therefore can never be realized in any specific single form. In the cosmos, in nature and through human nature *becoming* has created for itself manifold, indeed, infinite, forms that embody its expressions. These enveloping forms are made from the stuff of space-time and gravity, taken from the external physical world, and are shaped from within by means of becoming's intensive components and energies. In the first stage of de-actualization, the external enveloping and embodying elements are separated from the inward intensive elements, and (as happens in any death process), the envelopes, abandoned by their intensive life, eventually will fall apart and disintegrate.

2. De-actualization releases also the specific quanta of invested formative force, realized in the formation process of the enveloping forms. This formative energy has two sides. One is directed to the external physical world, taking up its elements, giving

them shape and consistency, restricting, compressing and adapting actualization to the conditions of the physical world; on the other side it is receptive to the intensive forces emanating from the self-actualizing and self-expressing event.

3. When the enveloping forms fall apart and the formative forces are released, de-actualization continues to liberate also the intensive energy that motivates actualization in the first place, leading the whole natural, cognitive or social process of expression and actualization on the physical plane. This intensive energy is the impulse that connects the incarnating event to the formative forces and physical adaptation. It is "The thread upon which the spider descends" (and re-ascends), as D&G call it in *What is Philosophy?* When this intensity is released from its investment in the formative forces and in the formed substances, the virtual core of the event is released from its attachments to its envelopes, and all the forces involved in its incarnation process are returned to the virtual, immanent plane of consistency.

Seen in this way, de-actualization as the first step of singularization, is also *a reversal of actualization*, sending the spider upward, opening a line of flight for its release. This first part is crucial as it actualizes de-actualization, individuates de-individuation, personalizes impersonalization, etc., by mastering the release process from its embodiment in fixed forms. De-actualization is that part of actualization that enables it to reverse the direction of it embodiment in the forces of space-time and gravity. Actualization *freed* from gravity, space and time (as reversed actualization), will thereafter be able to operate in a free manner on the plane of virtual immanence and consistency and *actualize the event of becoming* in its pure virtual flowing stream, and then re-actualize it again in new and improved forms.

Virtual actualization and Re-actualization

When the power of actualization is released and is fully singularized (individualized), it becomes operative in its own virtual element, and thus becomes virtual actualization. It now has two main functions: one operates on the plane of immanence itself in developing pure virtual activity and creativity in the fields of non-organic life forces, and the second operates from the plane of immanence downwards to individualize and actualize its re-actualizations, re-incarnations and re-embodiments in new envelopes, habits and forms.

Virtual actualization

The function of virtual actualization is to take over the de-actualized and reversed elements and forces, released from the event's actualization, in order to virtualize and actualize them on the plane of immanence. As we saw, de-actualization releases *three* components, created by the event for a former embodiment: these are, a. formed physical, cognitive and social substances and structures, b. formative-structuring forces, and c. intense energy of embodiment. Therefore, *first*, virtual actualization begins its activity with the released and virtualized material contents, chaotized structural elements, potentized fragments of the enveloping sheaths of expression of all sorts; these material elements are used as humus soil, rich in long lasting decaying physical elements that function as a virtualized "compost" on the plane of immanence, constantly nourishing the virtual, "Cosmo-dynamic" husbandry. *Second,* it takes hold of becoming's formative force, that being now freed from gravity's pull and the restricting and hardening physical conditions and adaptation pressures, can open up and give itself entirely to the virtual formative forces, radiating from the whole cosmic firmament. This freed formative force, is used to receive and return purely virtual formative forces and it is actualized in the formation process of purely virtual creation. *Third,* virtual actualization's main operative power is made of the reversed intensive energy invested in formed embodiment. Now, it is wholly distilled and purified, and becomes infinitely intensified, and it directs itself to motivate and actualize the virtual creative actualizations and individuations of the event itself on its own plane, with its own forces of non-organic life.

Let us look somewhat closer at the first stage of virtual actualization. In this stage it assimilates to its virtual activity and virtual plane the specific de-actualized components that were formerly incarnated in concrete and realized forms and substances. If we restrict ourselves only to embodied human life and cognition, these freed components were embodied and expressed through our instincts, desires, habits, feelings and emotions, sense perceptions, sensations and percepts, mental pictures, representations, concepts, and ideas. The event's last embodiment, now de-actualized, has endowed each component with a specific "signature," which it then carries with it as a virtualized quality and intensity on the plane of immanence. This is an essential distinction, not usually made. De-actualization, or counter-effectuation, is not "release for release's sake." It is creative, because it *carries over* from the physical to the virtual world, individuated, differentiated, personalized *signatures* that could have been gained only in and through actualization and embodiment in the physical world. On the given, cosmic plane of infinity they are *not* to be found, since they cannot be created and actualized there. As a matter of fact, individuation and actualization in form, is only possible in the physical world, because its unique adaptive space-time-gravity pressure, narrows down and weakens the otherwise intense cosmic expansive drive of the virtual. Again, singularization means exactly this: **capturing individuation and actualization, that can only take shape as such in the physical world, de-actualizing it, virtualizing its essential intensive potency, and applying the newly gained power to the virtual event of becoming**. Infinite cosmic becoming, remaining purely in itself, without going through individualized human becoming, would have never harbored individuation. But cosmic becoming includes natural and human evolution; it individuates itself in nature and then reaches individual self-consciousness in and through the human, reversing individuation, liberating it and actualizing it as a wholly new and creative stream of cosmic becoming.

This is a remarkable metamorphosis and transubstantiation that we ourselves go through when we singularize consciously the very forces that have individuated us unconsciously through natural evolution. Now we are reversing and individuating them self-consciously. In this stage we experience how we become newborn again ourselves, together with so many life forms and creatures. Here we may begin to understand St. Paul's insight, when he exclaims that the whole of creation and its creatures are in labor's agony, waiting for the coming of true humanity, suffering under the forces of the judgment of form, which creation took on itself not for its own sake, but with hope, in order to let the human become the firstborn of creation to achieve liberation of individuated self-consciousness. This is because in and through the human stream of becoming, creation as a whole will pass through its individuation and be cosmically redeemed from God's judgment of form through God's grace of freedom. It is not a personal human being alone that is de-actualized, released, and virtualized, but together with her all the elements of the released event of becoming are liberated as well. In individualized, self-de-actualization, though it must take place *through* a personal form of embodiment, we can truly experience how cosmic becoming becomes different. In this way human becoming makes a real, substantial, *difference* in the universe, and contributes a new creative impulse to cosmic becoming.

In reality it is always the whole of humanity, nature, and also the cosmos, that undergoes this individualized becoming. We realize now not only to what extent we are constricted, even unto death-like entombment, in confining forms, but the more we feel ourselves at home in the new real, cosmic, free life, the more we understand the creative potentials inherent in each single virtualized and signatured desire, emotion, and thought we ever embodied and incarnated in our physical life. Because we experience all our individuations, singularized and actualized on the plane of infinity, as creative seeds of new cosmic creation. On the virtual plane of immanence the released elements and stuffs of our life begin to breathe anew, expand, grow and become part of the streams of cosmic becoming, and find their associations and partnerships in the swarms and streams of cosmic networks and multitudes of events and beings.

And the same is done with the de-actualized and released formative forces and intensive energies of former embodiments. The more all the power invested in physical adaptations and actualizations is virtualized, the more it is distilled and separated from its concrete embodiments and can begin to virtually actualize its forces for themselves, the more it becomes here, on the plane of consistency, what can be called "the event of virtual actualization." This event—which will be described in detail in my forthcoming book on Cognitive Yoga—reveals the essence of the power of singularization, which can de-actualize, actualize, and re-actualize everything and everywhere in many different worlds. Here we individualize and singularize the power of actualization of non-organic life at its cosmic source and become creative on the plane of infinity. (This virtual actualization results among various new virtual creations also in the formation of a "body without organs," a virtual body formed outside the organic body as its virtual-cosmic correlate).

Re-actualization

Working on the plane of immanence, with fully singularized forces of the event of virtual actualization, we master the capacity to perform re-actualizations, re-incarnations and re-territorializations on any level and in any field of embodied matter, life, cognition and social life. We can do so for the first time consciously (as all given embodiments are done by nature, that is, unconsciously for us) and therefore we can learn how to create ever better forms and bodies for embodying the event. We learn how to create increasingly more delicate and subtle "resurrection" bodies, envelopes and sheaths of expression that will allow more transparency for its real essence, installing in them intensive, qualitative and physical improvements to allow for better conductivity of the event's radiation in the physical world. This can be achieved because we can re-actualize while transmitting consciously some potentized, rarefied, virtualized substantial intensities from the event of virtual actualization and infuse them into the formative forces and processes that shape its envelopes. We shape it from within the formative forces and processes of its incarnation, gaining more and more strength and primacy over against the selective, adaptive pressures of the physical world. We learn consciously to do what evolution has really been doing unconsciously for billions of years, as we saw above in chapter one: we learn "to grow young" with each new re-actualization and re-in-carnation of the event.

The process of re-actualization as a whole includes these stages:

a) We have at our disposal de-actualized, extracted, signatured quanta of components, formative and intensive forces from a formerly realized, embodied event.

b) The de-actualization practice has already prepared suitable soil for its healthy growth on the plane of immanence. We *plant* it there and let it be filled with new streams of non-organic life that saturates it with fresh energy of pure becoming.

c) When it has grown enough, we *harvest* it in a very special way. We divide it into two parts, as in a cell's mitosis, but use only *one half* of the genetic virtual potential. We keep the other half attached to its virtual source and in this way we reserve and *preserve* it in an undifferentiated, embryonic, virtual state, like a virtual stem cell.

d) We then take the second half and press it in and down to form a compact, consistent, virtually intensive force. We *contract and condense* its energy until it can be transformed into thought-full content. The power that motivates compression and condensation on the plane of immanence can only be described as a power emanating from an act of certain renouncement. This condensation can only take place if we renounce an option to remain forever in a seemingly blissful virtual existence. But a different sort of bliss results from opting for creative re-actualization, because we experience what it means to make an individual difference in the universe, and also because there is nothing more joyful and satisfying then a self-actualization accomplished in order to supply other embodied life forms with new "lines of flight" needed for creation's future redemption. When virtual compression and condensation is accomplished, thought is embodied and born, because in reality what we call human thought is but renounced and hence condensed cosmic life. We do this while continuing to keep the other half in a state of virtual potentiality.

e) Next, we ex-press the newly conceived thought-full substance and push it through the veil or network of memory (the brain-bound network of subtle energy that underlies the faculty of memory), until it is taken hold of and reflected back through the hardware of the brain. Then if the operation is successful, we may await its reappearance "on the other side" of our consciousness as a realized "concept" that is emancipated from the virtual, and received by our ordinary thinking.

f) Then we *individualize* and *articulate* it through an intended connection that we established to a particular sense perception, bodily sensation or action, create an intentional mental picture of the object or deed, and name it appropriately, in order to finally:

g) Shape and keep it in our memory as an embodied and individual *representation*. However, this representation will be distinctly different from all our former and unconsciously created representations, because in this case we consciously retain for a thread that links it to *a reserved potential* by means of which it remains attached to its virtual source. This means that the new representation will be different, because it carries in it a re-mark of a real difference between the virtual and actual and therefore will provide an improved envelop of cognitive or social embodiment, with new elements and possibilities. For example, it will be, pictorially speaking, more pliable, flexible, and more open for variations and transformations. This means the cognitive or social forms created in this way, will also be much more easily discarded when their function is over, because the virtual that they actualize and never forget, will remind us of the vastly richer variations and differentiations also available to express this event, when changing circumstances demand it, and so on.

The whole process of re-actualization is done consciously, without losing sight and forgetting the potential of the reserved event. We attach and tab the potential of the reserved event to the re-actualization process, but keep it strictly separated from it, maintaining its existence on a higher floor or field of consistency. We let it hover above and accompany the whole re-actualization process as a cloud of the unborn, without letting it deplete its reserved virtual force in physical realization. The consummated re-actualization process and the reserved virtual force are like two identical twins that have been divided at their virtual conception like Castor and Pollux and separated in the moment of conceptual or social realization. But as all twins do they retain a secret connection through a quantum-like non-locality and remote affiliation that continues to resonate and vibrate, also from afar, disjunctively, after the earthly birth and cosmic death of Castor.

We operate here with a capacity that allows us to "remember" non-locally and non-timely the event and its virtual origin in our virtualized memory during the whole series of these stages of re-actualization and realization, in the same manner—only *consciously*—that is done by all natural organisms. Take a tree, a crab, or a mammal. It "remembers" not to totally exhaust its living forces, not to kill itself immediately in incarnating, growing and maturing. It knows how to condense at each stage of growth only a part of its total living forces to form bark, exoskeleton or skeleton, but always retains a reservoir of unrealized, original, undifferentiated, stem life forces in its living core, in order to continue to grow, reproduce and multiply.

Admittedly, D&G's is a difficult cognitive operation and therefore also often misunderstood. The goal of de-actualization and de-territorialization is not mere "liberation" or "resistance." These operations are often usurped for political and social reasons as acts of wild and chaotic guerrilla warfare with philosophical means. But D&G insisted that what is liberated here is precisely the highly productive and creative power of actualization and creation, the power responsible for all realizations, manifestations and materializations. The importance of D&G's project—and this is why they constitute an ideal site for our project of *Event in Philosophy*—is that they generate a creative power not merely to "think" but to singularize: they help us produce and condense the actualization of the power of actualization outside the organic body, as well as making the event of its becoming into a vibrant and unforgettable concept. Therefore they have fashioned a much-needed constructive site that can operate in the middle of the creative process. In this virtual middle we singularize both movements of de-actualization and re-actualization, as well as the actualization of the virtual event itself.

It is important to emphasize again the difference between realization and re-actualization. Re-actualization makes realization possible, but is free from it. Without re-actualization there is no conscious realization. But realization that is oblivious to the actualization process forgets the *originary event* and therefore annihilates life without residue. It is killed in a rigid, dead form, regardless of the level on which the realization takes place, whether psychological, intellectual, or social-political. However, conscious re-actualization isn't realized in realization. Re-actualization always keeps enough "Pollux" energy in reserve, as reservoir of the virtual event. It is actualized in each act of realization only as far as the enveloping process extends, but without losing touch with the event. From one de-actualization to the next, through virtual actualization, the new re-actualization will have formed for itself new clothing with new sheaths of realization and manifestation, and it will make them more delicate, flexible and transparent, more transitory as well, like the colored petals of the flower or the butterfly's wings. Because they are so intensely living they are also so welcoming to death. But death means a new chance for fresh and creative de-actualizations, virtual actualizations, and ever better re-actualizations.

Body Without Organs

Now we can mention one of the first virtual creations actualized through the virtual actualization work on the plane of immanence. (The process of virtual actualization outside of the organic body will be described in detail in my book on Cognitive Yoga). D&G call it "a body without organs." This body is woven of non-organic formative forces and intensive energy of actualization that have been liberated from their embodiment and virtually actualized on the plane of immanence. This body is needed in order to carry a consistent kernel of our whole human personality (also called "singularized individuality") from physical cognitive and social life into the virtual and back. Cognitive Yoga's practice will utilize this body as a movable "shuttle," to facilitate a new way of "breathing" in and out, to exit the organic body and return to it at will. The creative agent itself, a singularized human stream of becoming, builds this body for itself, for the simple reason that it must embody itself in the new world of non-organic life as well as in the exit and reentry processes. This body functions in the virtual world as our organic body does in the physical world, as a vehicle of situated life, locomotion, and as tool for creative work. In the same way, the *singularized individuality,* as virtual agent, must be securely embodied in such solidly actualized virtuality made of tightly and consistently woven non-organic fibers. The virtual body without organs will enable a singularized individuality to breathe pure life outside the earthly, organic, atmosphere and learn how to creatively move, breath and circulate between earth and cosmos, carrying nourishing fruits and life giving seeds from each side to the other. For this singularized individuality must learn to become and function with full self-consciousness at home here below and well as there above, and in and through the exhalation and inhalation journeys in-between. Thus the virtual actualization of a body without organs, the virtual carrier of a singularized (individualized) human-cosmic stream of becoming, is the first stage in the creation of that future becoming, which D&G also called: the new earth and new people.

Deleuze-Foucault (D&F)

> "What Event, what law do they obey, these mutations that suddenly decide that things are no longer perceived, described, expressed, characterized, classified, and known in the same way, and that it is no longer wealth, living beings, and discourse that are presented to

> knowledge in the interstices of words or through their transparency, but Beings radically different from them?"
> (Michel Foucault, *The Order of Things*)

A synthesis without which the fundamental hybridization of D&G described above would not have been so fruitful took place in the meeting of Deleuze and Foucault (D&F). From our point of view, this meeting was arguably one of the most important creative meetings between thinkers that shaped poststructural philosophy in the 2nd half of the 20th century. However, only a few indications can be given here concerning such a momentous spiritual event. What is the function of Foucault's thinking in the composition of Event and Philosophy? It serves as yeast, fermentation substance, enlivening and awakening, contributing powerful igniting energy and capacity, which are absolutely vital for this project. D&F ignites the site and brings the "great work" into real internal combustion and motion. It fires the compositional event into actual motion, so it can start its becoming process directly from the middle of its becoming. If we place Foucault's thinking in this context, it will require us constantly to *think* the place in which the differentiating difference operates and actualizes itself, as the middle in becoming and as the middle of becoming. And this becoming becomes then our own becoming: Both "I" and "World" begin to vibrate together and to become through each other something different; and this takes place in this ungraspable, unseen and unheard of middle.

The work of Foucault is an expression of unbroken loyalty to an effort to grasp this middle and be grasped by it; a middle so un-mediated, so fluctuating and fleeting, which creates a borderline and traverses it immediately, and constantly reverses itself inside-out and outside-in. In reversing and turning, it constitutes limits, thresholds, folds, selves and worlds. It becomes like so many "I"s and so many "worlds," assemblages of micro "seeings" and "speakings" of countless beings. Such multiples, hybrids and *becomings* mark zones of indiscernibility, in which any differentiating and multiplying difference is actualized. Self-moving and self-reversing, this activity self-divides and multiplies thresholds, which constitute identities and dissolve them as they arise, using the borderline as a site in which thinking becomes "the fiery reversing sword," whose operations both block and open the gates of paradise (the actualizing of virtual actuality).

Of course, and this is unavoidable, many have lost their path in the dangerous zones of the borderline and threshold. Some fall prey to the well-hidden traps and landmines that densely populate these uncharted, new fields, in-between and on both sides of these borders. But Foucault's thinking can guide us if we follow him faithfully, because he is the new "cartographer," discoverer of new lands. Dangerous indeed are these territories, but also the most fertile and rich in precious stones and minerals of becoming, and, above all, human consciousness must venture now to discover and inhabit them. Modern consciousness, therefore, truly aware of itself and its perilous task, can learn how to venture forth into these new territories, to go in and come out, not only whole and safe, but also much enriched and matured. Then it can use the forces acquired in the crossing process of the thresholds in order to live and work in the open fields of non-organic, cosmic life (and this is where, as we saw above, D&G help us to construct a site of composition).

From the various essential concepts that Deleuze found and elaborated in the thought of Foucault, I will single out only two: *discourse* and *subjectivation*. (The third essential concept of Foucault, the concept of "power" and the various manners of its historical and social implementations, was also taken up by Deleuze, but we will not discuss it here). Their contribution to D&G will be understood if we realize that they mark the contours and limits of the new territory that D&G sought to open and navigate. Foucault was charting new paths that will lead D&G to develop the capacity to liberate the power of actualization from its various actualizations (as shown above).

Now the concept of *discourse* delimits a path that leads to the operation of de-actualization in the field of the thinking of language. It opens a way to free the thinking of language from the accustomed language of thinking, while the concept of *subjectivation* marks the direction in which thinking must proceed, in order to appropriately problematize the actualization of "subject-hood" on the plane of immanence. The concept of *discourse* wishes to name what actualizes language and what is actualized through language, without itself being realized; what is reserved as pure "event" of language after we purified it from its entanglements and relations to language itself as grammatical and semantic body of expression, a speaking and knowing subject, and the object intended through language. That is, Foucault is seeking a language that does not exhaust itself in being spoken and known after being spoken, a language that is not given in what is said about this or that. He is seeking a language that is not subjective or a property of a speaking subject, nor objective, phenomenologically speaking, as intending an intended object.

Furthermore, discourse is not actually given in phenomenological, grammatical or semantic elements or structures that ex-press it in language. The uncovering of the structures of discourse is actualized by performing de-actualization and de-territorialization of the real uses of language; uncovering the various structures, levels and ramifications of discourse, the multiple variations and metamorphoses that are expressed in language. It is practiced "in order to return the propositions into their pure distribution [....] to find again the source of their emergent becoming... to locate their happening as event." (*The Archaeology of Knowledge*) This is a search for the thinking of *language as an event*, which is a becoming taking new place and time in the middle that antedates the reflective self-knowledge of a speaking subject, to the same extent that it "happens before" the use of language in order to name and intend an object of reference. Before language turns into a signifying and expressive system of signs, it is experienced as universal discourse, or an "archive," which being itself invisible and non-personal, makes visibility (audibility) and personality possible.

Discourse takes place, therefore, in the open space, wholly external to the speaking subject and to subject-object relations constituted by language. Here Foucault delineated a 'line of flight' that Deleuze understood quite well because he had been hunting for it throughout his early works on Kant, Spinoza, Hume, Bergson, and Nietzsche. Deleuze was searching for the real difference between univocal being, the virtual, the actual and their realizations. The concept of discourse created by Foucault *eventalizes* language (to use Foucault's expression), or *virtualizes* it, because it differentiates between what is purely unthought and un-spoken in language—as event—from what is actualized and realized through language as meaning, reference, denotation and signification. Deleuze will use and expand this line of flight through his entire work. The bond that unites the two thinkers, creating the compositional dissonance we term D&F as an event of actual meeting, has its deep foundation in this process.

The concept *subjectivation* belongs to Foucault's later thinking. It was valuable for Deleuze because the concept of the "self" or "I" remained elusive and truly problematic for him as well. Foucault formulated the concept of subjectivation as a real problem of post-structuralism as follows:

> What must my "I" be, my "I" who is the thought that I am thinking, in order to be what I don't think, in order for my thinking to be what I am not? ... How is it possible that a being, that can easily be characterized by the fact that 'it has thoughts,' and is probably the only thinking being, has an essential and irrevocable relation to what is unthought-of? [...] What is unthought-of is not outside thinking but in the very heart of thought. (*The Order of Things*)

It will be difficult to find in contemporary philosophy a more accurate definition of the most problematic riddle of the middle. Foucault tried to find the historical roots of subjectivation in his late researches on *The History of Sexuality*. But the seeds are already present in his book *Madness and Civilization: a History of Insanity in the Age of Reason* (1972). It is in elaborating and transforming this concept that Deleuze expresses the becoming of D&F most pertinently in the last chapter of his book *Foucault*:

> Foucault never ceased placing inwardness under close scrutiny. But an inwardness deeper than any inner world, as the outside is farther than any external world...? [An inwardness] that is not different from externality, but precisely the inwardness of the outside...

The process by means of which subjectivity is produced is, according to Foucault, a real projective-geometrical inversion of the external-world-circumference outside in, creating an inner space, as a "fold" of the open, infinite world. Deleuze continues, "The inner as folding of the outside: it seems that this theme occupies Foucault through his entire work; an inner that is but fold of the outside, as if the ship is the fold of the ocean." For example, concerning the Renaissance's fool exiled out to the sea [the Flying Dutchman], Foucault says: he was placed in the inwardness of the outside, and vice versa.

"The ship (the self) is but a fold of the ocean (the universe)"! Such a thought is truly worthy of the event of modern philosophy. It is a powerful wakening and enlivening ferment in the construction work of Event and Philosophy. It leads us directly to the place in which thresholds become actualized and operative as complex chaoids of conflicting forces, perpetually and incessantly reversing not only, still externally speaking, all dimensions into each other, but more profoundly, reversing dimensionality as such. The "differentiating difference" becomes a "self" and "I" when the process of crossing the borderland of the threshold, problematizes our entire self-existence and self-consciousness. Such a problematic 'self' is the crucial gift offered to our work by assimilating D&F's hybrid into the site of construction.

Let us assimilate and individualize this "fold of self," the becoming subject, by actualizing it on the threshold. On the threshold, at the portal, it will take on the form of a sign of both warning and premonition, transfiguring *what was me* into this problem: how does metamorphosis become the event, in which the '*I think*' becomes *IT thinks in me*? Furthermore, how does *IT thinks me* become *IT thinks ITs self*? How is a "*self*" constituted, when *IT thinks me*, becomes "my" self while becoming *ITself*? The inverse correlation of the same problematic metamorphosis is this: what is revealed when that which was my "inner world" is unfolded, inverted inside out and dispersed and distributed in the external world? What comes to light when it is externally expanded and becomes visible in the open cosmic light? What is revealed when everything I came to be as my innermost subjectivity, and also as part of my unconscious becoming, appears in the light of "world memory"? What kind of a "world self" carries my inverted self in the open external world? Does the open cosmos re-member me when I am lost to my subjective inwardness? Do I shine forth with my full inward darkness when I am spread wide-open in the cosmic light?

For where does non-personal individuation take place and what new polarity of *self-and-world* is constituted in this event? How does this cosmic wholeness, complexity, and multiplicity sliced in twain, split and differentiated, polarize ITself into two halves, "my" self on the one hand and "the world" on the other, in a pre-individual individuation process? Finally, *what* and *who* becomes this new "human fold" which embodies and actualizes, as self-becoming world, as world becoming self, at this threshold? (In the next chapter on Event in Art, it will be suggested that this "threshold identity" is the first stage of artistic becoming).

Below we shall continue to compose together D&G with D&F, and add some more unique fermentation substances and potions extracted from Derrida, Badiou and Levinas, and add them to our composting process and mixture, as potentizing and energizing powers. We will intensify this alchemical process until the Event in Philosophy project grows into an independent virtual "Homunculus," ready to stir and awaken to full virtual life.

Jacques Derrida: The General Strategy of Deconstruction

Deconstruction, although it has become a popular term, should not be judged first and foremost as yet another "system" of thinking. Such an effort will be fruitless and misguided alike. Rather, we can approach it as a symptom of the radical change of human consciousness in the second part of the 20th century. This evolution expresses itself in all fields of thought, art, science and social life and a proper understanding of Derrida's endeavor can be fruitful only if we approach it in this way. In order to understand this evolution better, we must come to terms with one of its central aspects. This aspect is pertinent to the understanding of the current transformation of thinking. It can be described as follows.

Since the 1930s, and more forcefully since the middle of the 20th century, a highly creative force has been streaming into human becoming. I have variously referred to this in-steaming force both here and in my other books, as "the Event" of our age. The same can be characterized also as an influx, that introduces into our being as a whole, and also specifically into thinking, a new creative impulse and force. It is a power of growth, transformation, metamorphosis, in short, a powerful impulsion that on the one hand destabilizes and shatters existing structures, and on the other, enables wholly new horizons of creativity.

This "creative impulse" causes powerful organic-psychological and cognitive transformations of the human as we know it. It accelerates and intensifies an already operating and advancing separation process between the physical body and its organic life- and formative forces. This separation is felt strongly by creative people, if not always consciously, yet as an inner soul experience and unconscious existential *Stimmung* (atmosphere, ambience, or mood). It was a profound source of human creativity in the 20th century and is turning into something even more potent in this century.

In the 3rd chapter of my book, *The New Experience of the Supersensible*, I described this process and indicated some results of the drastic and far-reaching transformation of the whole human constitution that is causes:

> This loosening of the organic life forces from their physical foundations, from the head down to the heart, has a most decisive influence on the soul and ego organizations of the whole human constitution. It causes a real division of lower and upper man. We can discover an excarnating-dispersing effect on the upper part of the ego together with a parallel, too strong incarnation in the lower pole. This growing polarization brings about a far-reaching transformation of the whole human organization and with it an enhanced separation of the soul forces.

This separation enables thinkers, artists, scientists and inventors, to perform hitherto unheard of creative and intellectual operations. Derrida's thought is another symptom of this transformation. Unconsciously for him, the above-described process of separation and polarization is taking place. It is making itself felt in the depth of his soul in such a way that, in his conscious mind and thinking, he creates a complicated "thinking machine" that he names "deconstruction," whose operation is nurtured and fired by this unconscious separation and polarization of the human constitution. In his conscious thought he gives expression to the unconscious universal evolutionary process described above and creates a noteworthy reflection of its elements and dynamics.

The following passage is taken from Derrida's definition of the essence of deconstruction:

> What interested me then [when Derrida was writing "La dissémination" and "La mythologie blanche," in the late '60s], was a 'general economy,' a kind of general strategy of deconstruction. The latter goal is to avoid both simply neutralizing the binary oppositions of metaphysics and simply residing within the closed field of these oppositions, thereby confirming it. Therefore we must proceed using a double gesture, according to a unity that is both systematic and in itself divided, a double writing, that is, a writing that is in and of itself multiple... On the one hand, we must traverse a phase of overturning. To do justice to this necessity is to recognize that in a classical philosophical opposition we are not dealing with the peaceful coexistence of a vis-à-vis, but rather with a violent hierarchy... To deconstruct that opposition, first of all, is to overturn the hierarchy at a given moment.... [And on the other hand,] we must also mark the interval between inversion, which brings low what was high, and the irruptive emergence of a new

"concept," a concept that can no longer be, and never could be, included in the previous regime.

(*Positions*, p. 41-42)

Let us observe more closely Derrida's outline of his "general strategy" or "general economy" of deconstruction. It includes a general definition of any structural starting point as well as a threefold deconstructive apparatus (or machine) whose stated aim is creative, namely, to enable a creation of a new concept.

1. The starting point: A unity that is both systematic and in itself divided. ("In a classical philosophical opposition we are not dealing with the peaceful coexistence of a vis-à-vis, but rather with a violent hierarchy").

This structure will be transformed by using the following technique:

2. Deconstruct that opposition: Using a double gesture... a double writing... that is in and of itself multiple...
3a. Phase of overturning: overturns the hierarchy at a given moment.
3b. Mark the interval between inversion and
3c. The irruptive emergence of a new "concept."

Let us recapitulate Derrida's threefold stride (3a-c above): (1) we "overturn" the opposites of a given (violent) hierarchical structure. This means, "Bringing low what is high" (and vice versa). (2) While practicing this overturning we "mark the interval between inversion" and (3) grasp the "irruptive emergence" of a new "concept." This new concept "can no longer be, and never could be, included in the previous regime." It is a truly new concept. The whole procedure can also be understood in terms of a general economy of energy: its production, investment, release and transformation. Specifically, it aims to release and reapply a quantity of force invested in former, traditional and metaphysical thought creations. It trains us to unearth and discover old and fossilized underground thought treasures. This energy was embedded, calcified, in older "geological" strata of thought. Deconstruction's general economy strives to liberate this force in order to use it for the fresh creation of new concepts.

Essential to this economy or strategy is the realization that "in a classical philosophical opposition we are not dealing with the peaceful coexistence of a vis-à-vis, but rather with a violent hierarchy," that is, we start with the realization that a force was applied to condense and conserve hierarchical order and structure. The power used for this condensation is also the power of in-carnation, of de-virtualizing a stream of becoming, which has materialized and been engraved in social-cultural institutions and systems, through the powerful and often violent pressure of theological, logical, and political gravity. By analogy, the same force that fossilizes the cosmic life in the earth's vegetation and transforms it under pressure into coal and oil, works also in human thinking over millennia. It condensed cosmic thinking, as expression of non-organic life, into the organic and bodily cognitive foundations and their cultural and political correlates. Philosophically this is achieved by the use of classical hierarchical philosophical oppositions. Beginning in the Greek and continuing through medieval thinking, it was kept intact as habits of thought or 'truisms' that undergirded western religion, philosophy, metaphysics, and political and social life. It was preserved by means of the fact that it was violently invested in incarnated structures that led the human soul ever deeper into material existence.

The deconstructive threefold method outlined above is therefore a general strategy of liberation, devised to free this bound energy from its fundamental enslavement to traditional and classical hierarchical oppositions. This is done not in order to create new, perhaps differently named, hierarchical systems of opposition, but to free thinking altogether from the traditional *gravity* of thought, language and social structures. The goal is to open up new spaces of creativity and to let new concepts emerge, which are not the children of this structure.

The "overturning" maneuver is devised for this purpose. In the moment that the high becomes low and vice versa, a powerful energy shift takes place, the structure is broken in its middle, and a force held fast in the structure is suddenly released. A quantum of fossilized energy is made free, and can be put in the service of new creative moment. It is not a revolutionary and revisionist effort, that is, it doesn't wish to discover a more "primordial" or more essential hierarchical structure. As we saw above in Event in History, such an effort can only lead to a creation of a new hierarchy, a seemingly "new" but indeed as ancient as possible political regime, ruled variously by the "party," "state," or "market" and even "democracy," that only replaces the name of the old God of metaphysics, logic and religion, without changing its essence. This will only be illusory liberation, actually reinstituting the old hierarchy with vengeance under new names, while claiming their final demolishment.

Deconstruction, therefore, is not aiming to overturn Platonism while keeping its hierarchical values, but to use the overturning maneuver in order to transform consciousness and perception in the process. What really matters in the overturning is not the replacement of opposites into each other, but to "mark the interval" surfacing through the overturning. The interval—this should be grasped quite exactly—shows itself only in its emergent becoming, appearing through and between the overturning gesture and movement, on the one hand, and the "irruptive emergence of a new concept," on the other.

In Hegel, double negation is the dynamic soul and life force of the dialectic machinery. In contrast, fracturing and emergence are the innermost motors of the strategic and economic machinery of deconstruction. *Différance* is produced and set into operation as an interval that is virtually real; an active and productive *différance*. (Deleuze will also point out that this is not a method of *ordinary* differentiation, which he terms *differenciation* (sic), but a real differentiation, in which a virtual *différance* is produced and actualized. This brings about something truly novel that did not formerly exist as potentiality or possibility).

This interval is, therefore, a place of opening that is disclosed through a (violent) release of (violently invested) creative energy. When released, it is radiating through and in-between overturning and rupture. The released fossilized energy freed in and through the dynamic and double movement of overturning, gives itself as a free kinetic gift, free capital, and a pulse of energy that supplies the firing impulse that motivates an irruptive emergence of a wholly new concept. And what can this new concept become if it is thought through with the new power of creative becoming? It is first of all a new concept of the currently happening transformation of the whole human constitution, as described in the 3rd chapter of *The New Experience of the Supersensible*.

1. This loosening of the organic life forces from their physical foundations from the head down to the heart: This loosening overturns all binary, hierarchical oppositions between "head" (thinking) and "heart" (feeling), as well as between "head" and "limbs" (will, desire), and also between "heart" and "limbs," not only in classical philosophy, but also in life as a whole. It frees the middle heart-and breathing region from the dominance of the head (and limbs), and allows it to become a free peripherally creative middle—empty of any violent and hierarchical centering—in-between head and limbs. We realize for the first time that we become *a middle* if we grasp it not as physical location but as a "evental" place of constant overrunning of high into low, left into right—indeed of all dimensions into each other.

2. It causes a real division (interval, *différance*) of the lower and upper man. This division may of course become pathological if not lifted to full consciousness. But if we succeed in doing so, we discover a source of a powerful creative stream of becoming. This stream of freed life energy and formative, virtually differentiating forces, is emanating from the dividing zone—or zone of division—in the middle of our being. The middle becomes a source and wellspring of becoming, radiating living creative activity into the drying up and shriveling corpse of the brainy head

above, and inspiring stability and peace towards the volcanic metabolism below.

3. This growing polarization brings about a far-reaching transformation of the whole human organization: The cause of this transformation is the new emergent creative life force, liberated from its age-long enslavement, from the judgment of the God of form, in which it was violently captured and held fast by the increasingly materializing pull of gravity. It now facilitates a de-actualization and ex-carnation becoming process of the whole human constitution, as in the past it served its incarnation, individuation and actualization in the physical world. This happens because an evolutionary phase ended and a powerful in-volutionary phase of spiritualization began to operate through the "interval" opened—up and down—in the middle of the 20th century.

If studied in this manner, deconstruction may be experienced as yet another creative symptom, another differentiated actualization, of a wholly new virtual force operating unconsciously in humanity today. It demonstrates how, in and through the empty and dynamically self-overturning middle, through the interval—if we are fast enough to "mark" its constant activity—we may be able to co-create with an emergent becoming of new humanity and earth, which "can no longer be, and never could be, included in the previous regime."

Alain Badiou: The Subject of Truth

In order to "spark" Event and Philosophy into real life and becoming, it would be highly beneficial to "breathe in" the sharply delimited, mathematically defined, concept of the event of truth and the subject of truth formulated in Badiou's thinking in the last 30 years. This concept of event is a fiery thought production, a wonderfully chiseled thought work. It has something unique to offer to our composition, without which this crystalline force, so necessary for our construction site and so hard to find, may dissolve in too much air and water.

Let us first recapitulate some aspects of the Event in Philosophy discussed thus far. In the event "nothing took place... but everything changed" (Deleuze quotes from this passage of Peguy in three of his books). This means that nothing took place in the sense of external-extensive changes, but everything changed as a virtual "whole." However, let us not all-too-hastily, all-too-conveniently, confuse this difference with inner\outer, subjective\objective, essential\arbitrary distinctions. These common conceptual dualities actually mask and cover the virtually real—or actual—sense of the event. Let us say that what takes place in the event must be expressed thus: nothing took place except the place, that is, we must get rid of dualism to such an extent that the embodiment of the event in space, time, situation, bodies, will not in the least change its "immaculate virginity" and virtuality, that it will be thought through so consistently and immanently, without recourse or need to apply any form of dualism at all.

The virtual event, unfertile, "dead" (or better: unborn) in itself, has been actualized without changing any particular thing except everything. A new "truth" as Badiou would have said, has been revealed, that throws a wholly different light on the given state of affairs, on the situation, and this "truth event," or "event of truth," transforms our consciousness entirely. All our perceptions, affects, feelings, thoughts, memories, are changed, and we see the given situation wholly differently. The human seized by truth in the event knows herself, without knowing how it knows, to be wholly different from her-self, closer to herself than ever before and yet farther away from herself than ever.

As a matter of fact (as shall be described below in the last section on Levinas), life's horizontal advance from past to future, "the arrow of time," is a linear, past stream of time, constituted purely of judgment and doom (because it harbors only the finished work, the given). However, in order to grasp and comprehend the polar stream, the stream of pure becoming that flows from the future towards the past (the stream of freedom), it is paramount to first grasp the vertical active edge, a chasm, that cuts both in twain. The cut produced by an intensive differentiation, introduces a genuine break into the past stream of time. At the same time, it also separates the two time streams from each other, or better said, it uncovers and liberates the future time stream from its bondage to the past, and interlaces them into the conscious weaving of a 'new time.' This production of new time can also be variously called *end of time, apocalyptic time, the time of the event* or *the event of time.*

Now Badiou strives to think a (non) human becoming in and through the event. In order to do so, a real subject must be sharply differentiated from all human traits. Therefore, its formation must occur in a void that is situated in the proximity of the given situation as its own unrecognized "event horizon" or "black hole." There, according to Badiou, in what he terms the "event of truth," a real subject is created for the first time when truth penetrates through a given human material and situation. Now, a "truth process" will start as a struggle for continuous "fidelity" to the event and to the paradoxical making and preserving of new "evental" subjectivity: *subjectivity in truth*.

A real subject, according to Badiou, is created in such a truth event. The human animal, however, the biological-psychological-social being, is committed to the natural inertia and continuation of its daily life and its vested interests. Badiou designates a subject created by truth event "someone that enters into the composition of truth." This "someone" then struggles to hold onto this event, which breaks through all her existential parameters. The moment of true subjectivation is therefore also a moment of division, which breaks the human constitution, the situation and the inertial existential continuum. On one hand, the human animal strives to continue its "normal" (truth-less) life; on the other hand, it must relate to a subject of truth that is both beyond itself, but also in and through itself, "because the accidental trajectory of fidelity cuts through its [human-animal existence] penetrates its individual body, and marks him, from within time itself, in a moment of eternity." Furthermore, the constituted subject of truth, being penetrated by an eternal event of truth, and becoming truth's production, is caught in a cognitive paradox, because "it cannot know itself as capable of belonging at the same time to the continuation of its human-animal nature and to the accidental trajectory of truth; it cannot know itself as capable of becoming subject." (*Ethics*)

Badiou says, therefore, that the human must be able to:

(1) Be penetrated by truth in the event and to become a subject of truth;

(2) Be non-interestingly interested in fidelity to the event that divided it into subject and human-animal;

(3) Do this without being able to know itself—in ordinary consciousness—as a subject, because, as Badiou maintains, "to the extent that he enters into a composition of subject, to the extent that he becomes his own subjectivation, the 'someone' exists in unknowing of his own existence."

How can we think the being and becoming of this "someone" that should exist "in unknowing its own existence" (as subject of truth)? In other words, what kind of knowing\unknowing is this? What kind of consciousness should be able to grasp and support this unknowing through unbroken fidelity? No satisfactory "answers" can be given to such riddles, if we seek them *outside* the event of truth. As a matter of fact, it is Badiou's most significant contribution to contemporary thinking, that he adds such essential problems and riddles to fire its ongoing development. His *unsolved problems* become truly fruitful when they are properly activated in the location of the construction site of Event and Philosophy.

The subject of truth can only become a subject without knowing itself as a knowing subject. It cannot observe itself from outside and reflect itself as an object of cognition at the same time, because it belongs to the essence of subjectivation that it cannot be produced and observed as an objective and object-like process, and therefore it cannot be grasped by ordinary cognition. But now this subject, "incarnated" in the human-animal, should somehow be able *to convince itself* to transform its human-animal interests and use them in order to carry and support—through fidelity to truth—this subjective truth process. It would be a carrier of an unknown truth as "someone" that in turn should be carried and also nourished by the human-animal. It should develop what Badiou calls "subjective steadiness," which is "the engagement of your human-animal particularity with the steady continuation of the subject of truth... to let the invested inertia of the known serve the steadiness that is specific to the unknown." In other words, a subject of truth must create and establish a *counter-inertia* that works against the daily interests, habits, and existential inertia of the human-animal.

It must create "truth habits" or "truth instincts" which must not only take the place of ordinarily human habits and instincts, but also reverse them and use their energy to support and nourish truth processes of fidelity to the event. Eventually this opposite stream must become stronger than the given inertia, and create an opposite inertia, a steady consistency in the service of the "someone" composed as a subject of truth in the event. Badiou helps himself here with Lacan's formula: "Don't give up on your passion!" This means "don't give up the unconscious forces that constitute the subject of the unconsciousness." And Badiou transforms it into an ethical instruction: *Don't give up what we cannot know ourselves, that is, the process of fidelity to truth. Don't give up the unique manner in which the process of truth takes hold of you.* Become loyal to what cracks you open, divides and breaks your human animality, and this will found your new fidelity to truth, to an event that, for all purposes and for all the daily interests of the human-animal, has already vanished in the moment of its actualization.

It will not be too difficult to contend that Badiou's philosophical positions are consistently contradictory. He must maintain that—somehow—an un/known link must be established between the given human existence and the subjective process of truth, which is unconscious, or, better, super-conscious. But this means that the known must be somehow linked to and serve the unknown, the conscious must serve the unconscious or super-conscious. This requires, however:

Creation of linkage between the unknown and the known that must be both un-and- known. That is, it demands a certain knowing of what I don't know because I must build a bridge over the stream that divides the event of truth and its subject from the human-animal, and somewhere the two sides of the bridge must connect. The knowable side of the stream must be connected with the other, unknown, super-conscious, side; and "I" (but what kind of I?) should be "the one" that bridges them together. Yet this "I" cannot know the unknown, so how should I be able to mark and constitute a place on the other side, in which the bridge's head must be firmly anchored?

On the other hand, Badiou demands that on the given and known side of the bridge a site be built that will support, nourish and supply the whole bridging process from this side to the other. But this self-conscious human-animal side is, after all, using these resources for its own daily ends. How can I found a site in the given human-animal, in the known, that will really serve the truth process, since it is instinctively mobilized to serve its own human-animal interests?

In other words: It seems wholly paradoxical to maintain that I should be able to hold both sides of the stream together, because this means that "a middle" must be created. But this middle is precisely what cannot be constituted as long as on this side I can only rely on what I already forgot (which, as a matter of fact, I never could have know through my ordinary knowledge process in the first place), and on the other, I only *know* "my divided self" *unknowingly* as a subject of truth. To maintain that, nevertheless, I should be able to convert a significant amount of existential energy, forces and materials, subtract them from my human-animal interests, and channel them upstream to create this bridge, seems really impossible!

Badiou's solution to this problem constitutes the foundation of his Ethics: "Do whatever you can in order to persist steadily in that which disrupts your inertia. Grasp in your being that which grasped and broke you." Persist in this disruption, in perpetual overcoming of the temptation of forgetting the event and its newborn child, the "someone" as subject of truth. Use your human-animal forces in order to feed and support this unknown-subject. Badiou believes that it is possible to convert and transfer the human-animal forces and make them nourish a truth process: "Someone will feed it with his anxieties and excitements, someone else with his high stature and presence of mind, another with his excessive desires of control, another with his melancholy and another with his timidity."

The truths of science, art, politics and love are constituted and grounded in this manner both in our personal life and in human history. Truth breaks through accidently, and "someones," rare and unforeseen, manage to remain loyal to its truth process. And though Badiou will not approve of Levinas, nevertheless, the project of Event in Philosophy will only reach its required intensity if the truth of *the Other* will be experienced in the same way. Because this event breaks into my seemingly coherent, rational, *human all too human* habitual life, disrupts its inertial continuation, breaks my identity in two, and constitutes in me a new world, a new cosmos, wholly foreign, yet more intimate than all the intimacies of my known self.

Emmanuel Levinas: The Entry of the Other

What we need now, in order to complete the composition of the "holey five," is to introduce Levinas' major contribution to modern thinking. This is also a starting place for the becoming process of Cognitive Yoga, which actualizes an in-breathing and out-breathing of creative thinking. While D&G and D&F work from the threshold of difference more towards the pole of out breathing and de-actualization, greatly intensifying the actualization of virtualized thinking, Derrida's operation does not want to leave the zone of the threshold itself, helping us to deepen the actualization of *différance* and precisely delimiting the emergent point of new creativity. With Badiou and Levinas, we turn our attention to the other pole, and learn how to actualize the *in-breathing* process, taking in the event of truth (Badiou) and the event of the other (Levinas).

For the "I think," the unity of the thinking substance with itself is a given thought percept (with or without its dialectical development). We discussed above to what extent this unity offers the strongest resistance to deconstruction and de-actualization. Now in the meeting with the other person, it is much more difficult to uphold this resistance. (This is why

Husserl produced some of his most self-torturing texts when he tried to assimilate the other into the "I think"). The reason is that it is really hard to deny the existential and experiential fact that the other breaks through, and deeply disrupts and disjoints my seemingly unbroken inner unity with my-self as "I think." But again and again my ordinary consciousness rebels and hides this truth. When I stand over against the other in my daily cognition, I turn him immediately into another object of my intention; he becomes an object of perception and knowledge as any other object. Furthermore, even over against myself, I can assume an intentional position of external observer, and make "my self" into an objective percept or object. So long, therefore, as we keep to the phenomenological point of view, we can only describe external observation of states of affairs, processes of kinetic and formative change, and external or internal "lived experiences." These will both keep and enhance the unity of the "I think" as a block of individual identity (and the most common modern ground of universal dogmatism).

The only way out of this deadlock is the following. I can experience not only the fact that I am looking at you, but I can also have a rather fleeting and suppressed experience of "being looked at." I do not only perceive you, but I am also perceived by you. What would have happened to my "I think" continuum and fixed identity, if I could perceive the fact of my-being-perceived by you? What would I have to become if I could become conscious in the moment in which another being perceives and grasps me? What would be experienced "in me" in the moment that I grasp the fact of being grasped? (We should have asked Rousseau what it means, because he tried to hypnotize a lizard and was in turn hypnotized by it! Beckett should also be seriously consulted because he used Berkeley's formula: *I am perceived therefore I am,* in his struggle against Descartes' cogito).

In other words, what happens to my personal, inner, autonomous existence as a modern, self-enclosed subject, when I truly experience that I am grasped by a totally other being, by someone or something that is wholly, and irrevocably, other than my-self? In this case, I will be seized and taken away from myself by the reality of the real, the otherness of the other, the wholly strange and estranging pure event as such. The fact of me being grasped by wholly different otherness does not take place within ordinary consciousness, but only in a "different" consciousness, one that can become conscious of the event in the time of its happening. In ordinary consciousness the event is immediately "forgotten" (because after all it was never consciously perceived, only felt). I actually fall asleep in the meeting with the other, though this falling asleep occurs with great speed while my daily consciousness continues, seemingly uninterrupted.

The Other places unperceived "sleep," deep black unconsciousness, right inside the core of my radiant and wakeful daily consciousness, because it actualizes in the center of my unity with myself an unbridgeable otherness—its own foreign being. In doing so, it shatters and dissolves my self-continuum and unity, and opens a gap that as an existential abyss, or difference, may become the greatest help in the development of higher stages and states of consciousness. (This help is far more powerful than many inner "meditations" that only serve *to harden* the inner subjective ego-enclosure). Such an event, that lulls my waking consciousness to unconscious daily sleep, opens in me this "black hole" and makes it accessible during the day; it brings the night's otherwise oppressive and omnipotent power into "my" reach. It may help me—if I so desire—to lift up to consciousness the unconscious night that prevails all the time in my subconscious existence. Therefore, if I could perceive the presence that grasps me, or the "look" of the other while being looked at (as Sartre described with phenomenological accuracy), I would experience a total metamorphosis. In that singular moment, I would actually *become something else* (a stone, ant, flower, woman, star). For what is deep sleep but the cover that hides the fact that a third of the time of our earthly life is spent in becoming something entirely different from anything our daily self would have dared to know? To become a star or stone, for example, means to be grasped by their wholly different nature, becoming their non-human, strange world, joining the company of the beings and events that form and sustain them in the universe. I would then have to experience the most overwhelming anxiety and abyssal shock, because I would have entirely lost myself out there, becoming dismembered and scattered in the open, becoming a multitude, losing my center and finding infinitely diverging and multiplying universes taking the place of "my" self.

This means that the other is truly a liberator (and this is exactly what Levinas experienced so profoundly), a savior that frees me from bondage to my centralized, autonomous subjectivity. It frees me from my self and overtakes my place within myself, while, at the same time, allowing me to take a line of flight and to expand, decentralized and multifaceted, into the great open world outside. But then where would we find a "center" again? When I am lost outside myself in the vast universe, I can only recover a sense of self if I look back to remember the earth (remain true to the earth, as Nietzsche's Zarathustra said), to recognize what was "my" body, and experience that now another self is not only *inhabiting* the place of "my" destiny, but is becoming my fate. For Levinas, becoming aware of the "not I but the other in me," means always that if I wake up while being outside myself, I experience the total terror of my dismemberment, and can only find again a "center" when I find the other in me, investing me, incarnating in me, giving me the revelation of the true nature of reality. "He in me" shows me that the world's only center isn't an ego, and definitely not my "I think/feel/will therefore I am an ego," but my primeval subjugation to pure moral substance, a primordial, infinite obligation to the other (to use Levinas' expressions).

Of course, individuation, until now justified from a certain evolutionary point of view, suppresses this abyss of infinite love and moral obligation in our daily life, in order to consolidate our purely egotistical modern self-consciousness. But its time has come to an end as a leading, positive evolutionary impulse. Today we must find ways, conscious and healthy ways, to reopen what was closed, to free the captives, to actualize this openness, and to make ourselves a new body, a new fundament and ground. The new body will be a *reversed body*, the opposite of the body we had in the time of individuation: it will be another body, a virtual body without organs, the body of the other.

Let us look somewhat closer into this abyss whose desire surfaces in each moment in which I struggle to become conscious of the fact that I am grasped or looked at by another. Waking up while falling asleep means, on the cognitive level, experiencing something that can be compared to standing up while the body falls down. The impact of the other on me pushes me out of my body; when I am pushed out it takes my place like a sword pulled out of its scabbard, infinitely fast. I am now watching from outside myself the *other* taking hold of and inhabiting my vacant home, body, or center. This is a terribly unsettling and uneasy feeling. "I" find "my" self (precisely the terms that now designate what I have lost), in between two differences, two other-worldly existences and I am lost in both. For in what was once a bodily "me," there is now "you" (but as a wholly foreign, alien, unrecognized being), and outside, where I took flight, there is a world in which I become a multitude, a flowing stream of infinite *becomings*. In between such a "center" and "circumference," I find myself in this awakening, but in this moment the terror of *self loss* becomes totally unbearable (and hence I immediately fall asleep to protect my self from this double revelation and extinction).

Levinas firmly grasps that the ontological ground of subject-ivation is based on the fact that *you make me first* into a sub-ject at all. It therefore cannot be a subjective experience of an "ego" or "soul," because it is non-human, pre-individual, and primordial, and it comes before any self-consciousness starts to awaken. It is like the baby giving herself as subject to her mother's enveloping care and responsibility; it is the wholly naked, innocent existence *before existence*, being helplessly delivered onto a world, becoming the total carrier of the parents' presence. The primordial defenselessness of the infant, its in-fantile ontological situation (Agamben has finely developed this notion) is the very foundation of the pure assumption of responsibility for the other, because the other is nothing less than the giver of my self, my animator, who forever already in-spired my humanity and actualized it through this primordial inspiration. I learn to breathe as a newborn by breathing together with my mother's gentle, rhythmical breaths.

Now this experience cannot of course be gained by any phenomenological observation. It is *not* a Husserlian apperception of the other in my field of consciousness, and has no affinity with any "dialogue" with the other. (It is said that Levinas would become vehemently upset when his philosophy was compared to the dialogism of Buber). We are not speaking here about communication between *object-like* subjects, or about *mystical* experiences of harmony and unity with the other or the world. All these are purely *subjective* experiences, grounded in the ego, and perhaps *especially so* when the ego is striving to escape from itself through mystical ecstasy. Levinas requires from us, with his stern gaze, to listen to something altogether different: to the difference that constitutes the pure moral substance of obligation and duty, replacing Heidegger's "ontological difference" with a difference that grounds and antecedes both "self" and "world." I don't project a world as my phenomenological and existential horizon (Heidegger). Rather, it is the moral obligation to the other that pro-jects my sub-ject, and constitutes the foundation of my subjectivity, because it is the very moral ground of the world that meets "me" when the other takes hold of me. The other in me is not the guarantee of my subjective experience but the ground out of which my selfhood would emerge in the first place, primordially. On the contrary: the other in me *shatters my subjectivity entirely,* destroys my autonomy forever, and puts me into primordial responsibility and universal, wholly non-personal, ethical obligation towards him or her. He or she becomes flesh in me—incarnates in me—only because my flesh is from the very beginning made of the stuff of the logos, but a logos that is not the wisdom\beauty\might of the Greeks or Germans but a pure nakedness and helplessness of the dejected, persecuted other and each persecuted minority.

The other, as the new "human" center of selfhood and identity, would have always already imprinted, nay, in-carnated in "my" flesh, a revelation of the fact that my flesh, my body, is precisely not "mine" but ITS, HERS, HIS. This breaking through and opening of the "I," "ego" or "self" to the other gives me the power to stand outside my-self. This is the other in me as the primal, unconditional responsibility for the earth and all her children. Here Levinas finds the only place in which modern humanity, Cain's children, may begin to reverse the original reversal and reverse Cain's fundamental cynically questioning answer, *'Am I my brother's keeper?'* into a new affirmative answer, that will eventually become as primordial and constitutive of a future subjectivity as Cain's original deed is constitutive of our egoistic individuation: *Yes, I will become my brother's keeper!*

Chapter Four: The Event in Art

ARTISTIC WORK STRIVES to become "worthy of the event" and of the challenges of the 21st century, just like our sciences, history and philosophy. This becoming will be described in this chapter, bearing in mind that it will be a condensed introduction to rich fields of contemporary creativity. John Cage often experienced and described this new artistic experience: *"My favorite music is the music I haven't yet heard. I don't hear the music I write: I write in order to hear the music I have [not] yet heard...A music that transports the listener to the moment where he is [becoming]."*

This is a most exact description of the *Umstülpung*, or reversal and inversion inside out and outside in, of the physical and virtual states and processes of consciousness, from which the new artistic becoming begins. It must be noted, that Cage begins by saying what his favorite music is: the music he has not yet heard. This means nothing less than he is living in the midst of the process of becoming itself, and has already transformed his likes and dislikes, his pleasures, etc., into this new, future oriented, stream. Now this also makes it understandable why writing has a reversed significance for Cage. Writing is but a mirror that reflects becoming and consciously makes audible music that is not yet heard. The physical writing is for Cage only the means of becoming conscious of that, which, unconsciously (that is, beyond the threshold of ordinary consciousness) is striving to become conscious. Therefore, such music, when performed and listened to, is capable of bringing the listener into contact with his own stream of becoming, because he must listen to such music not with his ordinary, physically hardened cognition, perceptions, and judgment, but with a truly "open minded" cognition and heart.

This listening experience is described thus:

> We are living in a period in which many people have changed their mind about what the use of music is or could be for them. Something that doesn't speak or talk like a human being, that doesn't know its definition in the dictionary or its theory in the schools, that expressed itself simply by the fact of its vibrations. People paying attention to vibratory activity, not in reaction to a fixed ideal performance, but each time attentively to how it happens to be this time, not necessarily two times the same.
>
> (Cage, *Anarchic Harmony*)

My research of the Event in art, has found that there are seven elements, 'becomings,' or stages of metamorphosis that compose this Event in a non-linear and non-organic way:

1. Threshold identity
2. Initiation language
3. Sensation body
4. Vortex (Metamorphosis)
5. Consciousness crystal
6. Memory and destiny (Guardian of the Threshold)
7. History (Resurrection)

Threshold Identity

A work of art begins when it constitutes a wholly new artistic subjectivity on the threshold of modern consciousness and becoming. As we saw above, our time demands an awakening to full consciousness, cognition and creative thinking. The reason for this was also emphasized: the site of our truest becoming, the event of our time, takes place in the next, subtle, level of our life and existence. The event isn't happening in "another world," separate from ours, but we become conscious of it through changing our cognition and awareness *of the world*. Therefore we easily bypass it in our busy, down to earth daily life. But, on the other hand, our sciences and history, philosophy and arts express this subtle realm and actualize it. But also the meaning of this actualization is too often overlooked. Now between the two levels, the virtual and actual, there is naturally a threshold. And each field of thought and life, as we saw above, develops its own operations of passage, of crossing this threshold from here to there and back. And so does art.

On the one hand, art naturally aspires to grow to the next level of existence, to awaken a higher state of creative thinking and becoming, in order, on the other hand, to bring the fruits of this awakening and express them back in material life, in form. Art destructuralizes and virtualizes the given materials and penetrates through them into their hidden intensities and virtualities, in order to materialize again its experiences and insights. It goes through sensible matter, through colors, sounds, forms, tastes and touches, not into "another world," but precisely into the invisible side of this material world. It strives to perceive matter's virtual potentials, and re-express and re-actualize those potentials back in and through senses and bodies, but will do so through transformed sensibilities and bodily intensities and qualities. And the first stage of this process means also to produce a new "identity," one that is constituted on the threshold itself, because it is in the passage from one side to the other and back that art finds its grounding, its existential "security," which is pure movement, transition and reversal.

The practical difficulty experienced during this first stage is that this new identity can only be constituted while crossing the threshold itself. It is a practical art, not theoretical; it must be accomplished, and as a matter of fact, all great art, in all ages, however un- or semi- conscious, accomplishes this process of becoming, this constitution of new identity on the threshold. And the modern artist and the artistic work strive to make artistic work and becoming fully conscious. Yet, this striving is immediately confronted by the fact that the crossing is a constant unsettling, un-grounding, stormy irrupting movement of dispersing and dismembering processes of one's own self identity. For, we are all—and the artist in this respect is no exception—rooted so deeply in our physical life in an advanced technological civilization. Our cognition and instincts, emotions, needs and desires, are formed and hardened accordingly. The conscious crossing of the threshold must be, therefore, to begin with, an uncomfortable process, demanding quick adaptation to changing realities, flexibility of conduct and a readiness to view oneself and the world from ever-different perspectives. Conversely, when one has made a home in this process, reaching the other shore, she will have to find the inner strength to cool down and stabilize the intensive creative becoming streams of the virtual world of non-organic life.

This is the unique character of "threshold identity," that it is a twofold process, requiring diametrically opposite conducts and operations. First, in order to uproot one's roots in physical forms and habits, one must become secure in movement, transformation, change. But once on the other side, where changes and becoming are the rule, not the exception, she will have to bring about *stability*, order, and form into this world. It is, perhaps not entirely paradoxically, always the opposite capacities and operations that one has to develop in one level in order to get to the next, and vice versa; therefore, as one has to bring change and transformation to this world, one has to carry from here over the threshold some cooling, hardening and consolidating processes and forces, to create an island of stability in order to tame abundant becoming and transform it into a conscious creative process.

For this first stage of artistic becoming, the consolidation of threshold identity, one has to acquire the capacity to form a middle place, or zone, in which the two opposite movements will be both mastered as well as replaced according to need. On the one hand, this middle place must become a capacity to flow with and navigate the fluid and streaming process of change, which is the very nature of true artistic becoming. But it is one thing to flow with change, as it happens, and quite another to embody change, become change oneself to such an extent, that one can identify oneself with it as one's new identity; on the other hand, the opposite must become as true: one has to be able to stop the current, if and when one decides to do so, in full self-consciousness and voluntarily, and create forms in the flow, stable sites in the streaming becoming. And this is what after all art is all about, the flowing form, the spiraling vortex, the stream of chaotic becoming that isn't total randomness, but expresses new processes of becoming, which are as real and objective as our laws and forces in the physical world of fixed forms and finished objects. This place, or site, is therefore better called "an organ," a living, vibrating and pulsating organ. And factually, we all do have such an organ in our bodies; it is the heart with its blood circulation. For this reason we can also say: the whole becoming process of the artist and her work, with its seven stages mentioned above, is about creating a new, virtual, *heart and blood organ* in the "body without organs." As we shall see below, stage four, "the vortex of metamorphosis," will describe the formation process of this virtual heart in some detail. But it is important not to take the seven stages of metamorphosis in a linear fashion, as each of them includes the *whole process* of becoming in itself. So it is only natural to point out the fact that the creation of threshold identity and identity on the threshold, is already all about the laying of at least the foundation for the construction of the heart organ and its pulsating, rhythmic artistic functions.

The seed for this heart is constructed consciously in a place on a border that constantly shifts, separates, differentiates, but topologically also connects both sides. This can only be, from the point of view of ordinary cognition, an indiscernible and indeterminable middle zone, like what Foucault taught us. But nevertheless, it is not less determinable and discernible than our physical-organic heart and blood circulation. The only difference is—and this is a major difference of course—that the physical heart is *given to us* by nature and the virtual heart must be our free and conscious virtual *creation*, as part of a whole new body that we make for ourselves, for our new life, indeed, our "second life" in the real virtual world.

This middle is so hard to create because it is not a fixed and static form or situation, but a movable, muscle like activity; for, it is the heart that is always constituted at the beginning of any true creative becoming. Formed as it were from within and without, from both sides of the threshold, it connects them together rhythmically, and would create the necessary balance and harmony needed for the creative work to unfold, grow and bear fruit. It can do so because it connects the two sides of the threshold, the intellectual, finished form and state of daily cognition, and the state of constant becoming. The artist, as much as the scientist and philosopher and each person engaged in shaping history in daily social life, strives to create this conscious heart-felt bridge, to be able to peacefully exit the brain-bound cognition, delve consciously into the virtual non-organic life forces, take from there the inspiration for change, beauty, growth and healing, and re-enter the body and the world of fixed forms in order to transform it accordingly.

Such a threshold site and organ of becoming has been poetically described: "I begin to know myself. I don't exist...I am nothing. I will always be nothing. I cannot will anything. Moreover, all the dreams of the world are inside me...all lovers kissed within me, all idle wanderers slept on me for a while, all dejected rested for a moment on my shoulder," as Álvaro de

Campos (one of the becomings, metamorphoses, or incorporations of Fernando Pessoa) was bound to say.

In other words, actualizing threshold identity is an a priori construction work, a precondition for any work of art and yet it leads directly, deeply, into the very becoming process of the work as a whole. And the reason for this is that modern artists seek to realize an artistic work of becoming by *becoming a work* themselves. True artists have always experienced this mystery, where the "subjective" transformation of the being of the artist becomes objective, and where the "objective" piece of art in the making, becomes a living being, saturated with its own life and even soul and spirit. Often it happens as a gift, quite unconsciously to the artist. Today the age of consciousness is upon us and this means that this process must be raised to full consciousness and volition. Everything will depend, then, on our ability to practically realize this first stage of constituting a new, stable, harmonious identity on the shifting and unsecure threshold between two states of consciousness and being.

For this the artist strives to constitute a secure dwelling place on the threshold and *as a threshold* ("the place in me is pure space," writes the poet Razabi). She must however be ready to invest real heart-building forces in this work. They must be etherized, virtualized, taken out of the biological organ formation and carried to the virtual field. This is indeed a gift of one's own life, as any real work of art is. Therefore, *threshold identity* is the name given both to artist and work in the first stage of his—and the work's—becoming. The more this threshold identity is constituted (a further elaboration is beyond our scope here, and the reader is referred to my other books for more details), the more it becomes true that in the moment in which "I (subjectively) can speak no more," as Leonard Cohen sings, I can become someone for the first time "who is (objectively) spoken for."

That is to say, the artist becomes a new heart, a new speech, and therefore also a new language, because through the threshold process, and the beginning of the new heart formation, she is *spoken* as she is speaking; becoming a threshold between the ordinary mundane language and the new virtual language, a bridge between two worlds and existences that in our time must come together consciously more and more. She learns a whole new language, which can also be called: Initiation language.

Initiation language

This section will be the poorest, because, basically speaking, everything we said about the Event in science, history and philosophy, composes the vocabulary and syntax of the new artistic language. Besides, I would have to dedicate a whole book in itself even to the most rudimentary vocabulary of this language. Therefore, all we can do is to make some brief indications. This language may be said to have two aspects, *the scientific and the poetic*, or with their more traditional names: the plastic and the musical. Arguably, the most influential "founding fathers" of this new language are Deleuze and D&G (Deleuze practically covers every field of artistic creativity and experience in his books on Proust, Francis Bacon and Cinema, as do D&G in their book on Kafka and the magnificent chapter on art in *What is Philosophy?*). Essential to D&G's thinking as a whole, and this must also be pointed out in connection with art, is their unique ability to transform scientific and philosophical thinking into new artistic language, and what is more, to do so in a remarkably inventive way. When both aspects are combined, they reflect an unremitting endeavor in present art to create a modern initiation language that will be as projective/polar, chaotic, and intensive as Geometry, Mathematics, Physics and Biology became in the last century. This language creates for art new means of expression just as the mathematics of Riemann, Cantor, and Bourbaki, the geometry of János (Johann) Bolyai, Lobachevski and Mandelbrot, the theory of relativity, quantum mechanics and chaos theories created wholly new languages for the natural science in the course of the 20th century. This language realizes in art what the second scientific revolution, described in chapter one, has achieved for science.

First, it creates reversed geometries for artistic space by means of a metamorphosis of the ordinary spatial dimensions of art. Projective geometry (affine, polar, or synthetic) transforms the laws of Euclidean geometry, and at the same time contains it as specific case, a local realization, of a far vaster becoming of space(s). This is what, for instance, expressionism, surrealism, cubism and abstraction tried to grasp through modern painting and atonal, dodecaphonic, and serial composition in music. Furthermore, this language creates artistic laws of energy and heat, intensive "artistic thermodynamics," or, to use a more contemporary expression "non-linear dynamic modeling," which is inherently negatropic (anti-entropy) and autopoeitic (self-organizing) while at the same time world-open and differential. This is what, for example, Joseph Beuys tried to realize through his fat-felt-lead installations and actions, in order to invoke and express the creative warmth and fire forces and energies of art to kindle the flames of true initiation. Increasingly since the second half of the 20th century, art strives for the inwardly intensive, living and creative fire that builds and animates all matter with non-organic life and sentience.

Now a second, inwardly intensive formation of this initiation language comes from modern literature. Among many modern poets and writers, Franz Kafka cultivated this language in his way. (I can only point out now a single example, a single letter of this new artistic alphabet, because in reality the new artistic language is as diverse and rich as there are genuine artistic works that express it). Below I will list seven elements in Kafka's creation that articulate new signs, letters and words, through which his version of metamorphosis takes place on the threshold of modern consciousness. Kafka strives to demonstrate that "the force of the poem will be realized [only if] liberated from power... precisely by the retreat from and giving up of power." (Dan Miron). Kafka invents complex strategies designed to generate the tremendous force needed for *giving up power*; he invested his will to (non)power in order to retreat from power, and his tactics include a rich variety of highly refined maneuvers, creating disordered and distorted perspectives, negations and reversals, meanderings, confusing labyrinths. He mixes true, fake, false but especially uncertain, indiscernible displacements, all aiming to give an exact rendition of *the truth of failure*. This failure is actually the fullest success in giving up the forcing and enforcement (or "tracing" in D&G's language) of truth. This is the true reason why Kafka can give us such perfect descriptions of reversed metamorphosis. Kafka's intensive lifelong striving was, according to his friend Felix Weltsch, to "take part in wholeness."

And all his becomings and creations are expressed in the learned and practiced, conscious, almost *ritualistic failures to take hold of truth* as most of his contemporaries—and ours—do. He saw with clear sight at the beginning of the last century, how the three-headed beast of capitalism, fascism, and bolshevism was taking hold of humanity, in the west, in middle Europe and in the east (as described in chapter 2). Because he knew that truth is virtual (or spiritual) and hence infinite both in scope and power, it must, therefore, if it is to be socially and politically incarnated in healthy ways, do so only indicatively, differentially, facilitating, not reversing, freedom, equality and brotherhood. This is the reason why he *refuses* to take the path of forcing wholeness on social life as a "realization of one essential truth" through the power of religion, ideology, economy, or politics. This sophisticated "Jewish" strategy aimed to secure the survival of the weakest, is the strength of "minor and minority literature" (D&G). It is the reason why Kafka's truth emerges from *giving up* truth's power to subjugate and enslave, enforce and dictate. (This is also the reason for Kafka's ambivalence towards Zionism).

I believe it would be a mistake to read Kafka as if he means that true metamorphosis is impossible. He means only that *forcing* truth is in reality a reversal of truth, creating monstrous truths. Now this is an essential, political aspect of the new artistic language; it is as inwardly vigorous and even forceful, as it is externally resigned. Nothing would be more misleading than to interpret it as a language of neutrality or escape from social activism. It is, as a matter of fact, the very language of a new politics, but a politics that knows the secrets of virtual actualization, individuation, and differentiation of power, when it comes to social and human matters.

Therefore, Kafka refused to lie to himself and join those convinced that they could force truth but who are not conscious that they thereby only reverse and create economic, political and spiritual monsters (monstrous metamorphoses). But in so doing he realizes this failure in a positive sense, he virtualizes it, gives it voice and vocabulary on the plane of immanence. He immanentizes it, so to speak, and thereby implants in each failed becoming a virtualized force that nourishes seeds of true becoming. His reversed metamorphosis isn't just *a picture* of the evil reversal of modern politics, but it really becomes a fertile *soil* that supplies art with well-composted forces of growth and healing.

These are the seven stages of Kafka's failed becoming (according to Weltsch):

> 1. *Striving to take part in wholeness.* But wholeness is virtual, and not physically given in actualization.
>
> 2. *The defeat* (in realizing this striving). This is a victory, because the achievement of wholeness in the physical political sense is a monstrous disaster.
>
> 3. *Realizing the infinite distance between human and absolute.* This means, a sober purification of the illusions of unlimited *physical* personal, political, and spiritual power. But while uncovering the infinity of finitude it

prepares a complementary flip that the human, if traversed all the way through its finitude, through its death, reaches its infinite virtual essence in the universe.

4. *Being insecure*. This is the new security, the foundation in the abyss of finitudes, in so far as we are embodied and hence mortal, but which is the only guarantee of finding the possibility of virtual actualization beyond it.

5. *Being deserted, desolate*. This is the result of experiencing individuation all the way through, overcoming our inborn illusions about the possibility of "simple" or "natural" human relations. And its active, positive realization becomes a power of creating fully individualized interrelationships and communities.

6. *The role of women (hints of false redemption)*. Love is true redemption, but what is true love? Is it the naturally given love? Or the normative social and cultural image of love? No one who ever tried to really individualize it will underestimate Kafka's truthful, powerful resignation.

7. *Guilt (the fullness of failure)*. Being guilty and ashamed of it is the existential precondition for gaining a new power of virtualization; after all, if we haven't yet realized how far we have failed, where are we going to find the impulse for creative becoming?

Each of these stages, externally seeming to be a reversed, failed, metamorphosis, is inwardly, if so experienced, a well tested and individualized, source of courage. But this is something we have to *experience* with Kafka, as this is true for modern literature at its best. It is up to us to de-actualize it and to construct a line of flight, lay a plane of immanence, and actualize our own virtual becoming on this plane. If we do so, it will be inwardly felt as full of creative, positive intensity. Kafka's writing is altogether free from any trace of decadence; it is inwardly robust, healthy, and even joyful; his is a powerful new initiation language, and this is the reason why he sees so clearly through the modern mind and its delusions.

> Precisely because of this Kafka reached the highest level of resignation of power and prioritizing weakness instead of power; this is one of the fascinating elements of his personality and writing... in the hands of Kafka, weakness became the deepest and most penetrating critique of power. (Dan Miron)

But conversely, we must add: this weakness is far more importantly *a will to virtual power*, rather than a mere critique of external power. As initiation language, it feeds on death, lament, and helplessness, eschews complacent, pessimistic humanism, devours them, and increases its creative becoming.

Sensation body

An example of artistic becoming through the formation of an individualized sensation body (the sentient part of the "body without organs") is found in

Merleau-Ponty's fine observations in *The Visible and Invisible*. There he delimits the sensation body as a matrix of rhizomic webs, microscopic and semi-visible flows of becomings that the artist, spider-like, is spinning and weaving in and through him and the work. The artist expands and diversifies the limits of the personal sensitive aura or space of sensation, which in the case of an ordinary person does not reach far beyond the limits of his individual physical body. He is spinning and weaving in and through this expanded, sensitive, sensing body, and his work in progress is included in its growing and expanding radiation. The more his body grows and expands, the more he encompasses more of the work inside his body. He becomes himself a woven body of perceptions, sensations and sensitive, sensing vibrations; he is seized and is held fast by every percept perceived and any sensation sensed, and in being grasped and infused by what is sensed and grasped, he becomes spider and prey and cobweb, alternatively and continuously weaving between himself and the developing work. Any work of art at this stage can be felt as an inward force, because the artist's body of sensation has internalized the work by expanding his body over it. At this third stage of becoming, artist and work begin to grow together, expanding and shrinking in exchanging their identities while at the same time transforming each through the other: the artist becomes different the more the work grows, and vice versa, both metamorphosing into each other, to the extent the artist develops his sensation body according to the work's demand.

A keen observer of the body dynamics, psychology and physiology of artistic creation, will see that artistic work creates a new sensation body, which belongs to the same extent to artist and work. As a matter of fact, it is one indivisible body, in which both become at the same time. It is a virtual and invisible web-like scaffolding in which the following reversal takes place: the physically visible aspect of artist and work become increasingly invisible, and their invisible virtual aspect grows into physical visibility. This is a miraculous moment in the work of art in progress. Their essential becoming, woven together, is made visible. This happens after threshold identity and initiation language have been sufficiently actualized.

This cannot of course be objectively measured, but it can be clearly observed with developed virtual organs and capacities of perception. A moment comes in which the artistic becoming process, as a living organism, reaches the stage of maturity in which invisibility becomes consistent. This fine, living, branching sensation body (or sentient body) composes as it is composed, weaves as it is woven, creating the cobwebs of the work as its own individualized Aura and life-world. This body is a virtual complexity, diversity and multiplicity—the meridians of sensibility and sensation—a living realization of its artistic being and becoming. As matter is deposited and condensed out of pure energy, when the field's longing for, and surrender to, gravity, re-territorializes along the lines of force delimited on a virtual field, so does the work condense its invisible matrix to specific artistic matter. But we must remember that the event of actualization is *never* actualized in a body or a *body of work*. Out of the embryonic soup in the mother's womb, matter crystallizes. Cells are differentiated, organs are formed, bones are ossified, and arteries, nerves and fluids are materialized, just like paints, lines, colors, tunes, forms and shapes, but the event of condensation or materialization is not realized nor materialized. As pure, real, virtual actuality, it remains virtual, but sucks a quintessence of condensed potency into itself, tornado like, via each actualization; it harnesses it to its own field of forces, weaves a form of visibility around its invisible core, and gains increasingly an inner consistency at the same time that it increasingly make itself visible through its bodily envelopes.

Artistic becoming at the third stage means: creation of a non-organic and non-psychological sensation body on the plane of immanence, gathering multiplicities, webs and lines, to compose them as virtually consistent and embodied organized sensation. The real work as virtual event remains always invisible. But its appearance makes this invisibility visible; a work is the visibility of the invisible, as much as it is the invisibility of the visible. And as we shall see later, in the next stage (forming the vortex, the locus of metamorphosis), it is precisely this constant imperceptible interchange, replacement, and mutual reversal between the two that constitute an artistic becoming as real event.

The work's maturing physical or temporal appearance, either in space or in time, plastic or musical, is a composite (but inwardly always disjunctive, multiple) condensation of the sensation body's vibrating, gyrating, and spinning web of sensed perceptual substance, streams, flows and currents. The work condenses, brings to visibility, in a more or less fixed spatial or temporal form, this ever changing invisible composition, never fixed nor resting—an unfinished, virtual bodily work of becoming.

For example, take the description of the "text-textile" in Castel-Bloom's book *Textile*. There a formation of a *textual body without organs* is described, woven from the breaking down, detaching, diminishing life–and sensation streams of the characters. The "body of the text," the ancient and sacred art of life's weaving, is metamorphosed and de-actualized: from weaving the organic text (the living, organic body) it is transformed to the weaving of a non-organic sensation body. On the death bed of the mother (Amanda) her daughter (Lirit) remembers her words: "death is an un-weaving of a thread from the web of life." The whole novel is a grotesque, funny, painful, gruesome, lonely, and estranged description of getting old as a process of gradually loosening the threads of life, the threads of hope and meaning of organic sensation, feelings, desires, and emotional life. Therefore, everybody is dealing with threads, weaving, textiles, carpets, and cloths, in a desperate effort, an impossible and repeatedly frustrated desire, to be able again to weave oneself back into an embodiment in the "organic web of life." (Hence, for example, the security department requests Irad to produce a bulletproof anti-terror outfit made from spider threads stronger than steel). As a matter of fact, all the women in the novel are busily fighting death by artificially striving to prolong organic life through the constant re-weaving of *death's unweaving*. Thus they demonstrate that they have not yet understood the mother's future task, which Penelope understood so well. She reunites with Odysseus by unweaving her false marriage garments. But these women are unwittingly reversing Penelope's de-actualizations and so keep on weaving—day and night. They don't welcome the unweaving process of old age, as a blessed help for forming an individualized *body without organs* and they struggle to reject the falling-apart of life's natural-organic bodies and meanings. They are unable to accept death's gift of creative freedom, as an invitation, to learn the new art of non-organic life-and soul weaving. While longing for new life, they still don't accept the task of the conscious actualization of non-organic threads of life's new non-organic bodies, texts, sensations, or meanings. But their refusal is imbued with intense, though repressed, expectation for (essentially forever delayed and unreachable) redemption.

The novel ends when Lirit, the daughter, travels to a Kibbutz called "Ein-Kissufim-Ichud" (which means: Fountain of Longing), to meet an organic cotton grower (organic cotton symbolizes *the illusion* of a return to nature, whose living garments would—so she nostalgically hopes—unite again the impoverished life-and sensation bodies of humanity with nature's fountain of eternal life). On her way to the meeting she keeps remembering her mother's life long obsession, her constant weaving, connecting, and lacing, with which she was trying to overcome and heal disintegration, old age, decay and the inevitable death of the organic body. Gently and imperceptibly between the lines, appears a new, delicately transformed, stream of becoming; not a desperate fight to hold on and stop organic death, but an intimation of the free weaving of a non-organic matrix of new life, a post-dying text-textile, a sensation-body that senses and in sensing infuses non-organic life with fresh human and cosmic life forces and sensations, which is Art's essential task today. (A new mother's heritage is glimpsed, a future orientated tradition of the feminine, a reversal of ecological conservationism. Weaving is now transferred from the organic to the non-organic; a future art of cosmic weaving begins, of learning to weave new and vital life-bodies freed from physical-organic incarnation).

By now it may have become apparent that the becoming process of the artist's sensation body is in reality a new way to experience a bodily formative process. It must be seen as part of the artist's production of a body without organs. The sensation body is a contribution to its gradual formation, as were the two former stages. "Threshold identity" formed its initial heart-seed formation and "initiation language" added initial virtual speech organs and language forming processes. Now the sensation body formation adds an enveloping sheath to initial heart- and language formations. The body without organs is the work that is created in and through the whole artistic work of becoming, and as we shall see below, it is as much a part of the new body of the artist as part of the body of the work he creates.

The well known experiences of bodily and psychological disintegration and illness of the artist can be understood better through our interpretation rather than through a conventional psychoanalytic approach. Psychoanalysis simply possesses no notion of the evolution of consciousness in our age, and therefore is wholly oblivious of the virtualization process of the human body and its gradual timely separation from its organic foundations. But this virtualization is grounded in the disintegration of the physical body and the death processes that also accompany the building processes of the body without organs. However, the formation of the sensation body is not just coupled with death; it is death's twin or sister, for they belong together in one and the same bodily stream of becoming. Both witness the impersonal disintegration and deformation processes, which underlie any real artistic creativity.

The reason for this is that particularly in our age, in order to become a properly tuned and conductive instrument of cosmic forces of becoming, the artist's sensitive constitution registers the physical's body illness and death processes much more strongly than any other "normal" person. The artist cannot but feel intensely the physical body's death and virtualization process and the labor pains of an emerging embryonic invisible baby body. She is mothering a newly conceived creative offspring, fructified by the cosmos, and therefore she becomes aware to what extent the organic body almost dies when it is preparing itself to receive and conceive a cosmic influx of becoming. It becomes dismembered inwardly, and precisely in this way begins to transfer virtualized forces. Its dismembering creates a super-conductive substance that transfers inorganic fluidity, intensity and flexibility. A common threshold experience, the feeling that the ground falls away under one's feet, for example, is a result of this. For each of us, naturally, as we grow older, the body disturbingly falls apart all the time, the connectivity of tissues, the adhesion of muscles and the cohesion of the blood are hollowed out, dismembered inwardly, and so forth. But the artist experiences also, not always in clearest consciousness, how his solid and robust physical body begins to become a hive-like buzzing multiplicity, or better, a swarm-like assembly of independent and miniaturized cellular, molecular and microscopic particles and germs. The artist experiences how her own body doesn't belong to her anymore, how it seems to be contagioning, cohabitating and conspiring with others, becoming others' lives, invading any last possibility of privacy and autonomous inwardness. But in reality what is happening is that the body is being truly in-spired, hankering to liberate its bound organic life and spread it over and above its bodily organic limits, and is therefore often felt to be constantly threatening to expose and explode its organic form and also the subjective soul sensibilities and functions.

However, when understood from the other side of the threshold and not through psychoanalysis, this process is precisely the body's healthiest way of rejuvenation, of re-linking with terrestrial and cosmic forces of elemental life, through which the falling apart of organic connections between cells, tissues, organs and limbs, is compensated by a vigorous non-organic bodily formation processes and vitality. What is beginning to be felt, are many sensations that are unconscious in ordinary soul life. Microscopic, molecular, cellular, but also magnetic, electrical, chemical, mineral, and watery processes, radiations and emissions, are felt in the body in an increasingly disturbing way. Further, the more the falling apart of organic life is extenuated and psychologically registered, the more the body becomes sensitive to atmospheric forces and heat, climatic and meteorological fluctuations, and mineral, metal, and soil emanations. It begins to react in bizarre manners to plants, animals and humans, as well as to rural and urban environments, that grow into it as much as it grows into them. This, again, causes a loss of the given natural external and inward protections and boundaries, blurring the spatial limits of the body and subjective inner boundaries of the soul autonomy, mixing up perceptual contents and inner soul sensations with its external environment, which in itself becomes increasingly "ensouled."

In this manner Van Gogh feels how he is bodily penetrated and invaded by the intensity of colors, and to the extent that his organism disintegrates enough, it becomes transparent and conductive of the intensities of southern light, heat, soil and atmosphere. Normally inwardly experienced sensations like "hunger," "pain," "joy," or "sadness" become burning objective wounds, boundary breaking gates, through which flow moods of colors and tunes, which are, in themselves, concrete beings, relations, occurrences in nature and cosmos, now cohabitating the artist's body. There is nothing soft, soothing or comfortable about them, if we mean the ordinary, civilized and rather spoiled and lazy manner; they are often experienced as all too powerful intensities for body and soul, as objective, substantial, bodily attacks, inflicted against the organic will and natural self-perseverance instincts of bodily survival. Nature herself becomes expressive of elemental moods pulsating through metals, plants and animals, expressing their alien, non-human, existential reality and processes of becoming. The same happens with weather and temperature fluctuations, tempests or peaceful calm, earthquakes, rain, or hail, as in William Turner's paintings, when snow, avalanche, flood, and stormy ocean, become indistinguishable and expressive of the revelation of pure light, cosmically revealing its unseen intensity. But nature can express the unspeakable horror of human history as well, as in Munch's *Scream*, where she puts on crimson garments of the floods of blood spilled by generations past, filling and submerging the artist's body and soul with the bloody streams of the coming 20th century.

Here is Munch's testimony, as can be experienced through his sensation body:

> One evening I walked along a Hillside Path near Christiania together with two friends. It was a Time during which Life had ripped open my soul. The Sun went down, had dipped quickly below the Horizon. It was as if a Flaming Sword of Blood cut across the Firmament. The Air turned to Blood with cutting Veins of Flame. The Hillsides became a deep blue, the Fjord cut in a cold blue, yellow and red Colors. That shrill, bloody red, on the Road and the Railing. The Faces of my Friends became a garish yellow-white. I felt a huge Scream and I really heard a huge Scream. The Colors in Nature broke the Lines in Nature. The Lines and Colors quivered with Movement. These Vibrations of Light caused not only the Oscillation of my Eyes, my Ears were also affected and began to vibrate. So I actually heard a Scream. Then I painted The Scream.

Natural beings, historical events, catastrophes, the elements, all begin to express a new earth and humanity in becoming, and art and the artist's body and being become an expression of this becoming. Similarly, the new sensation or sentient body is a "sensitive-touch-body," in which sensations of touch becomes something much more intimate than ordinary physical touch. After all, any art must lie about external physical reality when it creates; it isn't satisfied with the given, physical, sense-perceptible visibilities. It strives for "higher reality" or even "higher truth." But what if truth itself—especially the highest truths—have been already thoroughly corrupted in human history, past and present? Where can he find a heaven of purity, of innocence? Perhaps we should travel "behind" the physical? But what if, when we actually do penetrate the "behind" of our human world, we find rather terrifying "heavens" of untruths? The densely accumulated results of humanity's corruption process of all truths? Because truth is already spoiled, necessarily contaminated in modern times (it cannot be protected any more by the gods and their servants on the earth). Any body of truths is already actualized as *a body of lies*; and these lying truths are deeply entrenched and embodied; they have become the flesh of our lives.

As untruths increasingly become given and inherited, the contaminated flesh, life-and sensation bodies of each human, the human body—and therefore art's body as well—becomes a location, a space, of the future human plague, a carrier of latter day's decomposing infection. Truthful artistic compositions therefore, necessarily start today with diseased decompositions. Their substance, these reversed truths (lies, violence, terror, evil) are actively embodied, and sprout, swarm and teem with infinitely vital viruses of de-composition (infinitely vital because nourished by truths). They are trans-fused and passed most intimately (as only true love can be infused) through human-to-human touching, via intimate mixing and brewing of "bodily fluids," tissues and fleshes, infected with our age's major illness. It constitutes a sacred place, a contaminated and sick grave of de-composition, in and through which Art alone can compose a new truth of de-composition and hence realize hope and healing.

This is described in a poem by Dori Manor, *Now it's HIV's Time* (in my literal translation):

> God's virus traverses the human soul,
> As underground cell it trespasses: on alert, risk group!
> The time yearns for the support of carriers of flesh and blood.

Vortex (metamorphosis)

This is the middle stage and central tool of the realization process of the work. It documents the becomings and metamorphoses of the artist's consciousness and being, and it takes place on the same plane of consistency and infinity on which also the philosopher, scientist and historian construct their creations. The real work's becomings and the real artist's becomings are happening on this plane. Our time is now ripe for a detailed research of this becoming through art, science, philosophy and history.

For the artist, the task is to create a point of reference for the work's concrete actualizations, which are real, consistent, condensations on the plane of infinity. This point is gradually enlarged through the previous stages of the work and is revealed increasingly—to virtual sight—as a virtual vortex of metamorphosis, an etheric cocoon building. As a physical image we can compare it to the insect's cocoon and the wonderfully mysterious metamorphosis happening inside it. However, this is only an external image. Therefore, when I speak below about constituting a vortex and/or becoming and metamorphosis, this must be understood as a *virtual* vortex of non-organic life forces and streams.

The construction of a vortex of metamorphosis takes place on three levels: first, *unconsciously* for us in the great work of cosmic art we call "nature," second, *semi-consciously* in any true artistic work, and third, as we strive to demonstrate here, it can be lifted one level higher and be practically realized as a *fully conscious* artistic becoming process.

What is the nature of this vortex formation? As we saw above, the artist progresses in the work by transforming his becoming through constituting a threshold identity, by developing a radically new, creative, initiation language, and by building a sensation body that unites his body with the planetary and cosmic streams of non-organic life and soul invested in the aura of the work. Up to the third stage we have to do with dominantly subjective-objective *projections* of the life and soul forces of the artist. However, from the fourth stage onward she becomes active and creative *in the stream itself* and must learn to completely detach her work from her own being.

For this purpose, she learns how to build in the *life stream outside* a permanent location or base, a "cocoon like" virtual construction. The function of this construction that is indeed shaped as a standing-flowing vortex in the external virtual world is to actualize constantly and repeatedly, by means of conscious volition, the virtual unity of tomb\womb through a fully conscious human-world interaction and composition. This is the meaning of artistic metamorphosis, the central secret of art's becoming. It is the process by means of which the released life forces, counter-effectuated and de-actualized and thus released from their bodily-organic basis, are projected from within outwardly by the artist's desire to expand his creative process beyond the limits of the organic body, and meet consciously the inwardly flowing cosmic stream. The vortex is formed in the external place in which the two streams—also called the microcosmic and macrocosmic streams—meet, struggle, fight, are fractured, transformed by each other and compose a creative, disjunctive synthesis, by means of which each can work in and through the other. (I have described this process at length in my book, *The New Experience of the Supersensible* as the "bridge construction work").

The meeting causes each stream to die partially through the impact of the other, and thus each stream is divided through the collision into two sub-streams, a dying stream in and through the other and a stream that retains its original life. One can even say (if we remember the inherent inadequacy of such analogies), that a kind of mutual fertilization occurs between the two, followed by a mutual mitosis event. Now, what takes place next is that an extract of the partial death of each is transmitted to the other, is exchanged between the two streams. It is taken up by the life of the other that then overcomes this death, and through this overcoming develops a new, inwardly enlivening force, which constitutes in each stream a joint product of the meeting of the two.

One part of the microcosmic, subjective, stream is killed by the cosmic stream, but then it is taken into it and is enlivened through it as well. It becomes in it a kind of compost and fertile life-giving seedbed active inside the cosmic stream itself. This is used for planting and fostering the cosmic seeds. On the other hand, the part of the cosmic life stream that dies into the microcosmic life stream, is then imbued by its life forces. In this way it can become a cosmic life power that can be active and creative inside the microcosmic life stream. This power is able to receive into itself the artist's released life and sensation streams of desires and intentions, purified through the meeting of the two streams, and gives them consistency and shape.

In other words, part of the artist's released natural life force is burned by the cosmic impact of the powerful non-organic life forces, and therefore it is dying a cosmic death, inside the cosmic life stream, and therefore it is also inwardly resurrected. We could also say that this part is dying as embodied organic-bodily and sensation life, and is resurrected as cosmic life and sensation. Its organic death process becomes alive, penetrated and transformed by undying, non-organic, cosmic life. The result is soil made of fertile ashes, thoroughly virtualized, that provide the cosmic forces with a suitable ground on which they can work, and which they can fertilize with their in-streaming inspiration seeds. This is the artist's donation and gift of a part of her released life and sensation forces. It supplies the cosmic stream—that kills and resurrects it—with virtualized human life-and sensation forces, by means of which the cosmic stream can work and shape from its side the work that is taking shape in the vortex of metamorphosis.

Now at the same time one part of the cosmic stream that dies when it meets the artist's life-and sensation streams is also resurrected back to life through the artist's microcosmic life forces. When the cosmic stream dies in the artist's life-stream, its death becomes life giving, and it provides it with germinating cosmic forces. The artist receives in this part of her released life forces, the remaining living part, a resurrected potency of cosmic life, which augments and intensifies her human creative forces, which flow as virtualizing force into the vortex of metamorphosis.

This is the real meaning of the vortex image: In the *in-between zone* the germinating forces bequeathed by the dying and resurrected cosmic life stream create one pole of consistency and intensity. It is the place in which the macrocosmic becomes immanent, actualized microcosmically as a centre for crystal and seed-like formative cosmic forces. The other pole is created through the resurrected dying forces of the artist's microcosmic life-and sensation forces, that become embedded in the cosmic stream and serve it as compost rich, fertile virtual soil. Both exchange mutually produced, mutually extracted, and mutually resurrected, creative potencies. Together they now work from both directions and sides, inversely reversing, inside out and outside in, their virtual re-actualizations and de-actualizations. This process as a whole is creating and shaping the vortex of metamorphosis, a cocoon like formation, in which artist and work achieve their first full stage of mutual mature becoming.

It is also the place in which the sensation body is fully transformed to become a bearer of the new formation and mixture of virtualized, non-organic, (cosmic) life in (earthly) death and (cosmic) death in (earthly) life. And inside the virtual cocoon formation of this vortex, a new essence of the work arises, a virtual "self," so to speak, a singularized "I" kernel of becoming is virtually actualized. This is actualized in a virtual interchange process by means of which:

1. The life and sensation forces that have been naturally embodied in the artist since conception de-actualize (or are brought into involution).

2. The non-personal, non-organic cosmic life stream is actualized and individualized in the place of the involving and de-personalizing microcosmic life-and sensation stream.

3. Both streams partially impregnate each other, as described above, to let their living-dying processes be embedded and nested in each other, weaving around and in-between the two poles, the virtual location called tomb\womb. This formation is a new

rhythmic system, in the centre of which a virtual heart-and lungs organ and processes are formed. In this place the wonder of metamorphosis occurs: a new, united, and yet objectively independent and separate, artist\work "singularity," is consolidated as the real being of the work of art. This resurrected quintessential heart-felt and heart-operative being is actualized in the streams of the open life-world, on the virtual plane of immanence. In the conscious becoming process of this vortex a new "singularity" is created as "immortal" artistic subjectivity through modern art. The world becomes a "heart," the artist becomes a rhythmically pulsating creative agent, and an objective contribution to the world's becoming process is actualized.

Now, an important note must be inserted here, in order to make clear the inner connections and mutual relations between the various functions of the four-legged project of the Event in science, history philosophy, and art. Each one of the four has a specific contribution to make to the whole of the virtual actualization project. Art's contribution, for example, through its seven stages of becoming, is the creation of this specific heart-and lung virtual formation, the vortex of metamorphosis. However, the vortex to be created here is based on philosophy's contribution, which is the development of a fully mature "body without organs" (see above, chapter 3), that has as its centre a virtual brain. This construction *cannot* be undertaken in the artistic process. It is the specifically unique function of philosophy as creative event. It can only be constructed through the work done in the Event in philosophy, whose main goal is to separate the software from hardware in the brain, and constitute an independent operative organ in the stream of virtual becoming. It is the unique task of the philosopher's becoming to construct this brain outside the organic body; and the artist's becoming must, therefore, be based on the philosopher's achievement.

The artist's operation has another, not less specific, contribution. The virtual, non-organic heart-vortex that it creates through each work of art, has the task of adding uniquely plastic and dynamic structuring capabilities and flexibilities to the body without organs and the virtual brain. This differentiated, mobile, dynamic, vortex formation of artistic becoming, is added inwardly to the formative forces of the virtual brain. As a matter of fact (this shall be demonstrated in detail in future work on artistic becoming), each artistic field contributes a unique virtual aspect of becoming to the structural formation of the virtual brain. (The full heart like vortex is completed first when each art has accomplished its formation work as part of the overall inner structuring and formatting of the body without organs).

For example, Virginia Woolf often testified about such a process of artistic becoming in her diaries, which transformed her life from relative, unstable, health to illness and from illness to creative process and inspiration. She offers many fine observations of what the artist seeks to learn to accomplish consciously and healthily. She believed that some of the symptoms of her illnesses were rather "mystical," because her brain stopped functioning for short periods. It stopped reacting to impressions. "It closes up; it becomes like a chrysalis." And then she felt a strange "flapping of wings" in her head and she lay down wholly paralyzed and numb. But on certain occasions, something "springs up" through her and she is "flooded by ideas." And she noted that this event occurred before she could regain control of her brain or seize her pen.

Here we have the main stages of what the artist, starting to be conscious of the work's becoming, wishes to lift to full—and healthy—consciousness. He also strives to "paralyze" his physical brain in order to press out its virtual brain; he strives, in other words (and as I wrote above, his work will be greatly furthered if he is supported by the creative endeavors of the scientist, philosopher and historian), to separate software from hardware first in his brain and then expand the separation process to the whole body, from the head down as well as upward. This separation is greatly supported by the formation process of the artistic vortex of metamorphosis. As art's becoming process needs philosophy's construction of a virtual brain, so the heart like, virtual vortex of artistic becoming, shapes and models inwardly the virtual brain by means of the above described forming process. Next, let us see how the process of artistic becoming is realized in a single painting.

Andrew Wyeth's World: the Becoming of Free Will
Comments on "Christina's World"

"I want the primitive effect you get when you bring together abstraction and the real." (Wyeth)

Christina's World seems to be laid out into the open, unfolding itself in a minimal interplay between figure, field and the houses on top of the hill that mark the horizon. The figure is leaning and projecting herself and her gaze expectantly towards the bigger building, as if she is in the middle of a movement intending forwards and upward. The direction of the unseen gaze, the leaning towards the house uphill, may be experienced also as inspiring hopeful expectancy, promising, perhaps, the fulfillment of a longing for home.

However, this world that spreads open between figure, field and longed-for home, hides an inner tension, which festers as a constant disturbance, the more we contemplate it. The more we do, the stronger becomes the tension and disturbance. The very openness of the scene begins to reverse itself. If we pay attention to this subtle, self-covering, reversal, it is experienced as circumvented on both sides: by the crawling figure on one hand, whose posture now suggests an arrested movement, a gesture that appears almost animalistic in its concentrated power of arrested striving, and by the gray, non-transparent horizon on the other, which may seem now as blocking the path forward and, at the same time, displacing and distancing the longed-for home.

The density and contraction of this limited world, contrasts starkly with its first appearance. It produces an almost physical effect that can halt our breath, even become psychologically oppressive, as we experience a tightly enclosing, narrowing, attraction between two movements, an arrested-expansion movement of the striving-crawling figure and the intensely contracted-opening movements of the promising hill-houses-horizon. An obstruction of the will, of movement and of freedom, is experienced, through which a non-spoken, inwardly intensifying, anguish is felt.

Wyeth's world unfolds and actualizes this tension in all stillness, in perfect calm, orchestrating—with minimalist means—an invisible, virtual cocoon, in which a secret metamorphosis is taking place. It is a process of arrhythmic, protracted, even spasm-like, folding and unfolding and refolding, spreading and gathering again of blocked intensities, an extractive operation enhancing itself inwardly, imperceptibly, and continuously. This appears to be the "primitive effect" Wyeth is striving for, when he brings together what he calls the abstract and the real (in the introductory quotation above). But his "primitive effect" is nothing primitive or given; its production is one of his secret creative techniques; he is a master of producing these "primitive" effects in the vortex, by means of his ability to discover, unearth and liberate an arrested movement, physically hindered will, preferably where it is not externally visible at all, where it is rather wholly absent: in stillness itself. As a matter of fact, the production of "real-ness" through Wyeth's style has nothing to do with a "realism" or "magic realism" commonly attributed to him. (The label "magical realism" could be used, but only if "magical" was connected to Novalis' wholly distinct meaning of the concept, as real, world-creative force).

Wyeth's real is the end result of a long process of productive realization (actualization), and he isn't interested at all in reproducing what we grasp as given with our senses and mind (realism), nor with imaginary effects that would conjure an improbable, haunting and strange world (magical realism). The haunting feeling that many experience with his paintings, its "strangeness" (Marc Rothko felt this to be Wyeth's main characteristic), is precisely the haunting of the real itself, which is produced by the painter. Reality must be first produced in the vortex of metamorphosis, and art's task is not a reproduction of the given, or even the uncovering of a hidden given, neither is it a fantastic creation of imaginary, non-existing worlds; but creation of a real new world, which, without art, would remain purely virtual. Art is as world creative as nature and the cosmos.

Wyeth's power is an operation, a technique, whose purpose is to extract a consistent essence out of the virtual process of metamorphosis, and the painting is made of this truly "magical" substance. He knows this just as well as Goethe. *Die and Become* is art's engine of production, namely, true becoming. We may glimpse an inkling concerning its production process if we pay closer attention to Wyeth's words: *"I search for the realness, the real feeling of a subject, all the texture around it...I always want to see the third dimension of something, not a frozen image in front of me...I want to come alive with the object."* Wyeth's creative production advances therefore in the following order of enhancement. First, the subject's feeling is released from its existential, realistic, personal, imprisonment (say, Christina's anguish). Second, its surrounding texture is gathered and condensed (the open as such must first be created, in order to be brought to bear, to embrace and carry the released anguish). Third, its third dimension is laid out as a plane of composition (a plane which dimensionalizes the arrested freed anguish, spreads it out into the impersonal, non-organic, real open field, where the subject can metamorphose itself, flow and spread its real wings into the open cosmos). Fourth, the artist has thereby truly become his object, or subject matter (he becomes Christina, takes on, releases, and transforms, her existential place, frees her will, so she can become her real self), and the mature work embodies both his becoming subject and the subject's growing into the free space of its real free becoming.

These four stages of enhancement are not subjective experiences of the painter, but stages of real becoming, the transformation and design of the reality of Christina's world, whose components are abstracted (extracted) from the subject, and made gradually more real, that is, consistently objective. Wyeth is after the potential, unseen, intensive will power, not the external will power. It flashes up when her arrested forward movement, on one hand, and opening contraction of her environment, on the other, are brought together in the vortex, as two opposing movements and intensities, where they interpenetrate and impregnate—without annihilating—each other.

This is Christina's world, which is indeed Wyeth's world: the world in which the miracle of metamorphosis is daily bread producing a virtual and nourishing stream of free will. This stream is not given—manifest or hidden—nor possible or imaginary, but what becomes through the larva of hidden, arrested, anguished will, when it opens to the cosmic forces of light. They are forming its wings of becoming and freeing it for infinite acceleration, actualized in and through the places in which it is most arrested. "I want to come alive with the object" means: to become the object's larval state, its embryonic potentiality, not its given, already actualized physical and psychological finished form and dead end. In this painting, one finds Christina's world, and by becoming her world, the artist wholly metamorphoses the energy of her larval anguish, her arrested will, in order to demonstrate its truth, its reality, its becoming. Wyeth let it become an eagle's flight, becoming an infinite free survey and homecoming, which embraces figure, field, houses and open horizon. (The same technique, applied not to the arrested bodily-physical will but to the arrested will power *of thinking and language*, is also one of Beckett's most creative poetic techniques. The affinities with another great contemporary American artist, Robert Wilson, in this regard, will have to be explored elsewhere).

Consciousness crystal

In stage 4 above central aspects of the vortex formation process were described in some detail. It was pointed out that the work's fourth, and central, stage of virtual becoming is expressed in the formation process of an artistic creative organ, shaped as virtual vortex, in which the work's as well as the artist's metamorphosis is taking place. When this stage is accomplished, work and artist have reached a first moment of objective confrontation. They stand over against each other, mutually expressing, reflecting, evaluating and re-forming each other. Now the becoming processes moves one step further into its fifth stage. What is formed now is yet another, *more inward,* organ, whose function is to lift the whole artistic becoming process into clear consciousness. What is described in this very book is being perceived and researched by means of *this organ* of consciousness.

The forming process of this organ condenses a clear and transparent space of stability and peacefulness, at the crossing point of the cosmic and microcosmic streams, through which the vortex is formed as described above. It is formed right where the two streams meet and fight each other. The place in which the vortex of metamorphosis is formed, is in the midst of mutually exchanging, replacing, constantly reversing, forces. There, through the mutual life and death and rebirth processes, a virtual organ of metamorphosis is formed. I wish to call it here the crystal of consciousness.

Now we will be able better to understand how this place becomes an organ of consciousness, if we pay attention to the following. Consciousness of this process is ignited in the moment in which the artistic becoming process comes together with the transformation of thinking accomplished through the work on the event in philosophy. Thereby a link is formed between philosophy's creation of a rudimentary virtual organ of thinking (the seed like formation from which later a full grown virtual brain will be constructed), and the heart-like formation described above. When, in this manner, virtual "brain" and virtual "heart" begin to communicate, the higher consciousness is ignited, and it becomes crystal clear, transparent, and self-conscious.

So we have to clearly differentiate these interrelated and mutually connected and nested virtual formations:

1. An embryonic formation of a "body without organs," with its initial virtual brain that has already been formed by the philosopher's transformation of thinking, described above in chapter three.

2. The artistic process of becoming, that reaches the stage in which a vortex of metamorphosis is formed.

3. An inner, virtual, mutual link is established between the two formations, in such a way that the brain becomes a heart and the heart becomes a brain. Both virtual formations complement each other, and each gives the other its unique characteristics and capacities. Virtualized thinking, by means of the link to the virtual heart, begins to sense and feel the inner qualities and nuances of the world of non-organic life, and the process of metamorphosis, described above, comes to light filled, virtual consciousness, by means of the virtual brain.

The crystal of consciousness is fully formed as an organ of virtual creative cognition, and becomes an innermost site of the whole virtual-artistic becoming and creativity. On the side of the *artistic work in progress*, this fifth stage of becoming completes the artistic work, endowing it with self-governing and independent life, consciousness and volition of its own. It has by now been fully born, in other words, both physically and virtually, and begins its external life in space and/or time of the real world. This means that at this (5th) stage, the work's forming process is firmly situated and positioned in the external world, physically as well as virtually; it is now truly objectified. The body of the work has now become a "new-born virtual body," whose virtual heart and blood circulation has become independent of the artist's mother womb, and is inwardly molded and differentiated by the forces of metamorphosis. Because this formative process has being completed, and the body of the work has reached independence, the formative forces are freed from their formative work. They can become fully conscious, because the vortex is completed and its rudimentary heart is functioning independently and therefore the new organ of cognition—the crystal of consciousness—can now be fully formed. The crystal of consciousness is located, therefore, in the external and open world, and it is formed there to the same extent that the physical work of art is already formed and completed in external physical space and time. There it functions as a new "cognitive-heart center" of the newly born body without (physical) organs. Each work of art has such an invisible, virtual, body without organs. And each artist develops unconsciously such an invisible body even if only rudimentarily, through his lifelong artistic becoming. Inside this body, made of individualized universal, non-organic, body, life and soul's formative forces, the organ of consciousness is formed and is now matured, and can be used to describe what we are describing here.

The consciousness crystal can also be compared to highly refined eye or lens of awareness, a brand new and rare virtual focusing instrument, which although still very primitive at the present stage of human evolution, nevertheless has a great future before it. It will become a new organ of virtual consciousness, by means of which we will see, think, navigate and become creative on the plane of immanence, in the open fields of the planetary and cosmic non-organic life. It will function in a fully voluntary and clear, self-conscious, manner and furthermore, will become an expression of humanity's stream of becoming. In the future—which starts already now—art's task will be to finely chisel and refine this lens. (Spinoza was refining it all his life and Fichte implanted it in the center of the "I" activity of self-actualization. Both are great teachers and forerunners of this consciousness eye crystal building). It is, indeed, a human-cosmic "third eye," a newly formed virtual pituitary gland, constructed in the center of the vortex in the innermost parts of the virtual brain, and linked directly with the heart organs and vortex of metamorphosis produced through art.

Let us observe more closely some aspects of the forming process of the crystal of consciousness. As pointed out above, this organ is the result of the conscious perception and thinking of the formation process of the crossing point of the microcosmic and macrocosmic streams described in stage 4 above. It is this "point" itself, which becomes a virtual "eye" or "lens." The substance needed to imbue its virtual formation with crystalline, transparent, consistency, in order to harden and consolidate its inward lens-like structure, is also produced there, at the point of crossing and metamorphosis, inside the center of the vortex itself. At this crossing point, at the bottom of the virtual vortex, as it were, a rhythmical distillation process is occurring all the time. It creates a residue of quintessential substance, a portion of virtualized dead leftovers that are taken up and filtered by the vortex's forces of metamorphosis. Now this distilled and refined substance is gradually infused with new non-organic formative forces streaming from the cosmos.

The crystal of consciousness is made out of a very remarkable stuff, art's produced quintessence, an extracted and finely virtualized mineral essence, made of transformed, resurrected, death. This finely virtualized dead ash is transformed through the process of metamorphosis and is infused by the new and fresh cosmic forces attracted and embedded in the work's metamorphosis. These forces endow it with its crystalline, etheric, transparency. It serves as a virtualized "dead" and hence transparent and selfless foundation, rhythmically purified from any human and subjective opacity. A quintessence is produced and formed from the artist's own released and virtualized life-and sensation bodies, transformed at the crossing process by the cosmic stream of non-organic life. Therefore, it can be said that consciousness crystal constitutes a place of quietness, stability and peace inside the otherwise flowing, conflicting currents of organic and non-organic life. Its formation process is nothing peaceful in itself; it is taking place in the midst of a rather violent and turbulent virtual vortex, in the center of which only slowly a quiet and peaceful island of clarity and transparency condenses itself.

In addition, the consciousness crystal functions as the cognitive part of a seed-like, embryonic heart-and lung organ, formed in the body without organs. This enables it to grasp, sense, monitor and distribute cosmic heart pulses, and transform them into conscious human pulses, registered through fully conscious human virtual cognition. That is, at the 5th stage, eventually a double "eye-heart" organ comes into being, linking virtual brain and heart together, in and through which virtual cognition becomes, at the same time, a conscious, co-creative cosmic-human pulse and circulation. This cognizes as it brings together, fuses, and harmonizes the streams of non-organic life that become *embodied in the work*, and the artist's organic life forces that are *released through the work*.

The crystal of consciousness can therefore both direct and consciously monitor the actualization process of the work. The work of art is revealed by its means, at this stage, in its truthful being. This being reveals itself now as a constant striving to become a place through which harmonizing and balancing of cosmic heart pulses and breathing rhythms with human pulse and breathing is happening as "event." This is where a "bridging" virtual singularization is accomplished, not only through thinking, but through the whole human body and constitution. A pulsating cognition, a breathing consciousness, which is at once "contemplative" and "operative," comes into being, "in-between," in the middle, and on both sides of the threshold, connecting, separating and connecting again, at will, the physical and virtual worlds. This is art's true meaning and becoming, to form and condense a health-giving, breathing, heart-felt place connecting the physical and virtual worlds, through which an interweaving can occur between cosmic heart's beating and human's heart's beating, which is after all what any real work of art is always (unconsciously) accomplishing.

What was achieved in stage 4 above (by the forces of metamorphosis in the virtual vortex), becomes here, in stage 5, a *detached organ* of perception and operation. This is the virtual correlate of the fact that now the physical work (the painting, sculpture, musical piece) becomes detached and independent from the artist; and, at the same time, the artist's being becomes to the same extent also detached and independent from the artist's ordinary personality. The work has now become an independent being, endowed with virtual-elemental consciousness, cognition, and volition of its own. It can therefore from now on both reflect on itself and on the artist, and on their relations; it can act, perceive and discern, and hence also write, document, code and decode virtually its own being and becoming; and moreover—this should be of great interest for the artist herself—the work, that has achieved mature objectivity and independence, is more than ready to voice its evaluations and judgments concerning the truth of the now almost finished work.

Memory and Destiny (Guardian of the Threshold)

Every truthful artistic becoming becomes as a matter of fact also a meeting with the "double" or "guardian of the threshold," because the work of art, which strives to realize an event of becoming, is embodying the artist's own metamorphosis and self-realization. Also the artist's virtual becoming, reaching an independent existence beside her ordinary personality, is asking for such an objective evaluation, nay, it is energetically promoting it. Therefore, at this 6th stage of becoming the work becomes the site of objective self-knowledge. The work, in this sense, often appears—necessarily we should add—also as disaster and failure, if we are objectively honest, because it is complete and must show us how incomplete our achievement is. Such a meeting is always dis-aster, or "blow of fate," because it actualizes true self-less-ness, which transforms this work and any work into really truthful work.

Ariel Hirschfeld offers a fine description of such a self-meeting with the double in his book, *Notes on Epiphany*. From his description we can estimate the full measure of the trauma involved in a full meeting with the guardian, because Hirschfeld makes clear that he meets only the very external appearance of his double. However, even at this preliminary stage, it is experienced as "terribly poisonous blade that penetrates me." The same is of course the subject of *The Picture of Dorian Gray*, which is perhaps the first modern literary expression, alongside with Dostoyevsky, of this meeting.

The cause of this self-meeting through the work is factual: it is an outcome of the fact that the work, after accomplishing stages 4 and 5, has become "finished work," that is, an independent being in the objective, external world. And its virtual being shows, in and through its very artistic virtual formation, a truthful reflection of the artist's being as the father-mother, as the creative source of this work. Meeting his *grown up work* becomes—as happens often with our children—"a day of judgment" for the artist. As Derrida pointed out in his remarkable confession (see the YouTube piece *The Fear of Writing*), the work's objective evaluation and judgment of the artist's being and work is a tremendous shock, indeed trauma, for the artist that is ready to apprehend it consciously.

For the artist this meeting will remain unconscious, unless she is energetically seeking it. If she truly desires it, however, this confrontation with the work's objective judgment will bring about a wholly new creative impulse, and the work, that was believed to be essentially complete, will open up to receive a radically new stream of becoming; the artist will be led again to question herself, to experience her "threshold identity" (stage 1) and start the artistic becoming process again on a *higher* and more selfless level of truthfulness. In the future such will be the necessary path of any artist that strives for true self-metamorphosis through her work. The disaster in the self-meeting through meeting the *being of the work* will constitute, then, a sacred moment, a blissful initiatory experience, and the work, that until now constituted only a vaguely general and largely unconscious artistic "event," will be transformed into an artistic "initiatory event."

When the crystalline organ of conscious virtual perception, that grasps the virtual being and virtual becoming of the work, is fully functioning (stage 5, above), the artist perceives directly to what extent he already shaped himself through the accomplished work. He can now see through the crystal of consciousness to what extent—in and through the work—he himself became an objective world-work; that what is reflected back to him isn't his known personal self at all, nor the external physical artistic work. Nothing known is reflected, and everything is revealed to be otherwise than expected and known. The guardian is a *"mirror"* in a totally different sense than the metaphor suggests in the ordinary use of the term, because what is "reflected" is a sum total of all the consequences of the work as viewed *virtually*, from a wholly objective point of view and reference. That is, we see the work's being in harmony or disharmony, its contribution for good or ill, to the real external life-world as a whole.

Therefore, each of the above described stages of becoming consist of a gradual intensification of the coming crisis, a step by step uncovering of a deep abyss, that is going to divide the artist's being by means of the work's being; a seemingly forever unbridgeable *différance* now opens up and down. It separates the personality from its new born virtual singular being, because now *the objectively appearing work* functions in unison with my *body without organs*, intervening and judging between me and my ordinary human self. The work demands objective answers to its evaluations and objections when this event, that remains largely unconscious in artistic creation, is lifted to full virtual consciousness and cognition. This happens when it is taken and posited in the vortex of metamorphosis (stage 4 above) and observed with the consciousness crystal's organ of perception. Then it becomes clear that this crisis can also become a blessed and graceful disaster, because the crisis itself (the trauma of which usually remains deeply unconscious and unhealed, as past trauma), is coming to meet the artist as an intimation of a virtual memory of the future. This unique virtual "memory," that can remember the future that *would have been* and the future that *can yet become*, opens the work and artist to a wholly new becoming process.

This new becoming will take place through an influx of fresh, truly virgin, cosmic non-organic life forces that begin to stream from the future towards the past. This reversal inside out, and outside in, of both work and artist, is revealed through the crystal of consciousness, that as a new, seed-like heart organ begins to operate also as a "organ of destiny." Such an organ can both perceive and shape future possibilities, based on true self-knowledge gained through confronting and assimilating consciously the wound of truth, as a powerful moral revelation. And the light that illuminates this crisis of self-knowledge, is the luminous truth revealed through the virtual being of the accomplished, past work. At such a moment the place of crisis and trauma is transformed into a place of *future birth*, through which healing and resurrection will become possible, provided that the artist is ready to rise up to this challenge and face her *self* through the work's objective judgment.

For example this was, rather consciously, the tragic source of the ingenious poetry of Avot Yeshurun:

> ... You asked me [in the interview] how does a man become Avot Yeshurun? The answer is: from the breaking down. I broke my mother and my father, I broke their home, and I broke down their peaceful nights. I broke their holidays, Shabbats; I broke their value-in-their-eyes. I broke their voice. I broke their language. I abhorred Yiddish; I took their holy language [Hebrew] into everyday life. I made their life a misery. I left the partnership.

What is vaguely perceived and generally called "destiny" or "fate," and taken to be only a dream of uncertainty and the play of blind chance in ordinary life, now becomes a conscious initiatory event. The guardian is the transformer and virtualizer, the awakener that lifts and transfers forgotten personal memory into more than ordinary fully conscious memory. It opens the gates of the future stream of time, which counter-effectuate the finished work. This is the meaning of this event that can become a healing event, but this healing is nothing psychological (in the sense of sealing the wound and lessening the pain) but purely virtual. It is a pure becoming of a wholly new, future creative work. Such an event of destiny is the objective gift of the work to its truthful lover and maker that by now has already become a really selfless artist. It makes possible new, fresh, artistic-moral decisions based on objective, evaluations and feedbacks. At this stage, artistic work gains for the first time its fully saturated moral meaning, as it becomes a member of cosmic life itself, sharing the fate of all free human creations in the universe.

At this sixth stage, art may become *real mystery art*, and the work becomes destiny, a future stream of resurrection. This was beautifully expressed by Jabès in his *Book of Questions*:

> Man enters the night
> As thread into needle,
> Through blessed gate
> Or gate saturated with blood,
> Through the most luminous of fissures.
> Becoming thread and needle
> Man enters the night
> As he who enters into himself.

History (Resurrection)

Now that the work is indeed really finished, and has finally been detached from its maker and become—after the "last touch"- a permanent part and parcel of the external world (physical and virtual), it has become "history," that is, it has been put in the grave of dead time. What becomes now of the work, what process of becoming it is going through after its final completeness, after its death? What does it go through in history, namely, in the grave of past and dead physical—and virtual -time? In his heroic and tragic, lonely fight against positivist 19th century historicism, Nietzsche emphasized the importance of "creative forgetfulness" as a necessary precondition for the development of real historical conscience and consciousness (see *The Use and Abuse of History*). This conscience and consciousness can serve as the foundation for a new kind of "memory," that remembers not the factual event that was realized physically and is recorded in history's archives, but the "event that could have been" and remained virtual. Or better said, this could be a memory that remembers the "virtual cloud or aura" that surrounded, and will eternally surround, the event realized in past history. Active, creative forgetfulness, Nietzsche writes, breaks and opens up the dead linear continuum of past historical time, imagined in human heads as a chain of causes and effects advancing from the past to the future, held and regulated as if by iron necessity. (Of course it has also been conceived in hindsight, and reconstructed "logically" by the historian, long after the facts have been registered, documented, and arranged in causal linearity). This creative *forgetting* is, therefore, active *remembering* of the virtual. It stops history's dream of linear causal continuation; it breaks the chain of causal command and logical necessity, which are the pure inventions of the intellectual mind utilized ever since Roman historians began to write distinct western history. Now we may enter the realm of actual virtuality and study "history" from Nietzsche's *perspectivist* outlook. (Here Deleuze's admonition resounds to turn our attention to the difference between "virtual" and "possible," in order to avoid any relapse to a phenomenological variation of possibilities).

Now what happens if the work and artist are put in the grave of history without going consciously through the preceding stages? What happens if the *Alp-Traum* (nightmare) of history doesn't become a conscious wakening event? In this case the final, 7th stage seals the total reversal of truth *into its opposite*. Its truth is finally reversed into an active falsehood without redemption.

But truthful art creates, through the moral forces gained in the fully conscious meeting with the guardian, the living memory of the future, planting a seed of new life in history's otherwise dead stream of time. However, the artist is fully embodied in his historical time; he is a true child of his age. He, more than any other, finds himself already placed (before he even started to create the work) and entangled in a complexity of non-human forces, desires, instincts and concepts that grow and thrive and are also nourished by his work. He cannot but become increasingly aware, if he seeks this awareness energetically, to what extent his work is nourished by the forces of the age and in turn nourishes them. In stage 6 it became painfully clear to what extent the work supports the spread of our current global social and cultural illness, to what extent my "good work" is already woven, before it started, with this stream of historical time.

Like Nietzsche, Rudolf Steiner also contended that modern, rational, historical science, born in the 19th century, and the historiography written in its vein is but a dream convinced that it is full wakefulness. It is actually an *Alptraum* precisely because it is so totally convinced that it is awake and illuminated, and therefore *lacks any desire* to wake up. The truth is that our modern historicism is the deepest historical sleep since humans began to experience history at all. But it is a dream and sleep from which we must awaken if we want to approach historical reality, in order to confront consciously and cooperate with the forces that shape creative historical becoming. (This is the reason why Joyce has Stephen Dedalus say in *Ulysses*: "History is a nightmare from which I am trying to awake").

In this sense it is true to say that the poetical, imaginative, dream of poets is closer to spiritual awakening than modern historicism. The artist is unconsciously closer to the forces that shape real history than the ordinary "science" of history. The artist's dream is at least a true dream because he dreams poetically inside the real formative historical forces, while the science of history, that believes itself fully awake to "real" history, is truly and tragically a dream that falls deeper and deeper into cognitive and moral oblivion the more modern humanity's disaster intensifies.

Charles Olson was also strongly aware of this fact. He could sense the difference between the Greek experience of "history" and what it became immediately thereafter through Roman history as the foundation of the western historical dream of rational enlightenment. He could sense how history already grew old in Rome, while in Greek times it still breathed the very last breath of its oriental youthfulness. With Herodotus, history was still *Istorin*, closely related to real becoming, animated by the last echo of the mythological goddess of time and destiny. How fast it becomes modern history through the beginning of the European, rationalistic, archival history in Thucydides! In Olson's reconstruction of Herodotus' *Istorin* we can feel how he strives to listen to the open spaces of formative, becoming time, the vastness of Aeons, the murmurings of virtual durations. But the archive's fullness of objective facts, of recorded, photographed historical "information" and infinite flood of "historical material," wholly charms, deludes and lulls us so easily into our modern cognitive sleep and moral forgetfulness.

But if the artist does not forget the future stream of awakening that he experienced through his wakeful trauma, he may join the life of the work also *after its final completion* and historical entombment. He joins the work's *future becoming* in its virtual after-life, through his own virtually actualized immortality. He remains true to its living being and goes with it down into history's grave of dead time, and he awakens together with the work's eternal being to greater and higher (because now morally motivated), creative consciousness. He increases his power of awakening inside the grave of modern scientific, rationalistic, technological dreams. He wakes up to deeper consciousness that is more real because it is virtualized, when she lets the historical fate of the work—as finished, dead work—become his and her profound Agon(y). He is then ready to take that path, the conscious crossing of the threshold that leads into the underworld, where he meets not only the souls of the deceased but where -today—he meets *the being* of historical death. This journey that became a tragedy in Late Greek times can become a source of most profound hope, such as it was still possible to experience in the ancient Greek mysteries before they closed their portals for the last time. Now it becomes possible again, but *not as an individual initiation experienced in isolated mystery centers,* but right here in the center of our modern civilization. This is an awakening event inside the breaking down and virtualization of all our known dimensions, as modern science and technology progress further and deeper beyond the threshold of organically embodied life. Dead time then breaks up and with it our conceptions of historical, linear, causal rationalism, or nihilistic indeterminism. History becomes a wholly different life, and in what otherwise appears only as a place without hope, a new stream of becoming emerges.

In the grave of time, of dead historical factuality, or nihilism, in the dream of (non)sense, the work is awaiting the resurrection of the artist's being and becoming. That work, which is all human work, and all the treasures of human history, is buried in dead past time, and can only be raised from the dead in the heart of truly awakened humans. As the Israeli poet Gilboa wrote: *The miracle is the cultivation and growth of reality.*

At the same historical place of time's breaking down, of time's death, if historical time is *not* raised into full consciousness, a radical reversal of the good is realized. But the same historical time, if raised to a fully artistic and moral consciousness of becoming, also turns into a place of real historical metamorphosis. It is precisely here that history becomes a source of future creative and joyful healing. We wished to demonstrate through the four branches of this research on the Event, in science, history, philosophy, and art, that we may be justified in harboring this hope and that such hope can be founded on a real resurrection experience. In a sense, everything woven together through these four branches is but a search for means to justify what perhaps only Levinas could have understood in the second half of the last century, namely, that "the essence of time is resurrection."

INDEX

CPSIA information can be obtained at www.ICGtesting.com
Printed in the USA
LVOW040726041012

301328LV00001B/166/P